The National Trust
RIVERS
OF BRITAIN

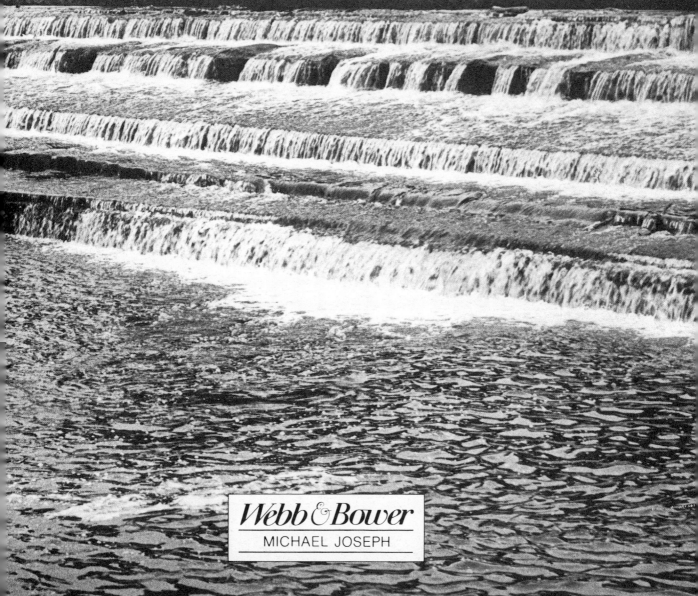

The National Trust
RIVERS
OF BRITAIN

Richard and Nina Muir

Webb & Bower

MICHAEL JOSEPH

Photography by Richard Muir (except where otherwise indicated)

First published in Great Britain 1986 by
Webb & Bower (Publishers) Limited
9 Colleton Crescent, Exeter, Devon EX2 4BY

in association with Michael Joseph Limited
27 Wright's Lane, London W8 5SL

Designed by Vic Giolitto

Production by Nick Facer

British Library Cataloguing in Publication Data

Muir, Richard, *1943-*
The National Trust book of the river.
1. Rivers—Great Britain
I. Title   II. Muir, Nina   III. National Trust
551.4′83′0941   GB1283

ISBN 0-86350-110-9

Typeset in Great Britain by P&M Typesetting Ltd, Exeter

Printed and bound in Great Britain by Hazell, Watson and Viney Limited,
A Member of the BPCC Group

Contents

Beginning at the Source

Rivers are rather like kittens or apples: almost everybody is fond of them. We open the first chapter with the words 'A river is water on the move' and my initial reaction to the suggestion of a book on this subject was the fear that, for all the charms of our rivers, it might be difficult to get beyond such statements of the obvious. In the event, however, we found that there were so many facets to the subject that the text is a little longer than at first intended.

We have not tried to write a conventional guide-book to Britain's rivers – it would be fraudulent to pretend that this could properly be achieved in a text of less than about a million words. Instead, we have sought to provide sufficient information about the origins, form, history, uses, abuses and ecology of the different rivers to allow the reader to gain an understanding of any chosen river. Also, rather than presenting a standard gazetteer, we have selected a large number of particularly relevant, inviting and rewarding sites and have set the information in panels which are placed beside the appropriate places in the text. So the book should be packed with ideas for places to visit although we hope that readers will appreciate that a book must fit within its covers, and only a fraction of the possible locations can be included. Nina, my wife, has written the chapters on the natural history and management of the river, while I have described the physical geography and human history of the river setting and provided the photographs.

In the distant past, each river-bank must have served as a routeway for settlers exploring the interior of this outpost of the European continent. Fortunately, many of these old ways endure as public footpaths. Yet in doing the essential fieldwork and photography I have been saddened to find how many of these paths have been lost to agriculture and to the more exclusive fishing clubs, so that the public is too often barred from lovely rivers like the Hampshire chalk streams or Herefordshire Wye. How different this is from the practice in Ireland, where good fishing and farming seem happily able to coexist with civilized attitudes to public access! It does seem to us, personally speaking, that the nurse, steelworker, shopkeeper and pensioner should have the same right to enjoy the pleasures of our riverside scene as the stockbroker or colonel. In several parts of the country the National Trust and the National Trust for Scotland provide access to some exceptional riverine sites, including some incomparably fine places, like Wicken Fen, a range of Lakeland settings and the Grey Mare's Tail Falls, while other organizations, like the county Naturalists' Trusts, the Wildfowl Trust and the RSPB also offer some wonderful attractions.

Most readers will have a river that is 'theirs'. Without a shadow of doubt we adopt the Nidd as 'ours'. For the first twenty years of my life I went to sleep lulled by the sound of water swishing down a great mill weir on the river, and my love of the countryside was nurtured in childhood walks along the valley. Subsequently I lived between the Dee and the Don in Aberdeen, close to the Liffey in Dublin and not far from the Granta or Cam. In contrast, the rivers of Nina's youth, like the Galana and the Pangani, supported hippo and crocodiles – but she too has fallen in love with the northern river which the old Britons called 'Brilliant Water'. Now for us it is a source of inexhaustible excitement that by the time this book is published we will be 'home' beside the Nidd, just a stone's throw from the old weir, with four different riverside walks on our doorstep and not a hostile Range Rover or bailiff to mar the scene.

Thanks in large measure are due to the late and greatly missed Robin Wright of the National Trust for his characteristic help and encouragement during the formative stages of this book; to the fish diseases research station at Huntingdon and to the NERC culture centre of algae and protozoa for assistance in obtaining photographic subjects, and to the various friends and experts whose brains we have picked for information about particular sites.

RICHARD MUIR
Great Shelford
March, 1985

Abbreviations Used in the Gazetteer

As many of the sites listed in the gazetteer of Chapters 7,8,9 and 11 are nature reserves, the managing bodies in charge have been indicated. As public access may be seasonal and limited it is best to get in touch before planning a visit for information and/or permission. It would be helpful to send a stamped addressed envelope with your request. Where necessary exact locations are indicated by the Ordnance Survey grid reference.

AWA	Anglian Water Authority
BBONT	Berkshire, Buckinghamshire and Oxfordshire Naturalists' Trust
BC	Borough Council (after a place name)
BNS	Bristol Naturalists' Trust
Cambient	Cambridgeshire and Isle of Ely Naturalists' Trust
CC	County Council (after a place name)
CCT	Cheshire Conservation Trust
CNT	Cornwall Naturalists' Trust
CTNC	Cumbria Trust for Nature Conservation
DC	District Council (after a place name)
DCCT	Durham County Conservation Trust
DNPA	Dartmoor National Park Authority
DNT	Derbyshire Naturalists' Trust
ENT	Essex Naturalists' Trust
FC	Forestry Commission
FSC	Field Studies Council
GTNC	Gloucestershire/Gwent Trust for Nature Conservation (depending on the text)
HIOWNT	Hampshire and Isle of Wight Naturalists' Trust
KTNC	Kent Trust for Nature Conservation
L and RTNC	Leicestershire and Rutland Trust for Nature Conservation
LTNC	Lancashire/Leicestershire Trust for Nature Conservation (depending on the text)
NCC	Nature Conservancy Council
NNT	Norfolk Naturalists' Trust
NT	National Trust
NTS	National Trust for Scotland
RDNHS	Ruislip and District Natural History Society
RSPB	Royal Society for the Protection of Birds
STNC	Salop/Suffolk/Sussex Trust for Nature Conservation (depending on the text)
WT	Wildfowl Trust

CHAPTER 1
Creating the Riverside Scene

A river is water on the move. As such, it becomes several other things as well. It is home to a fascinating array of plants and animals – indeed, one could argue that without rivers and streams there would be very little scope for life upon the land. But the river is also an efficient means of removing surplus water from the ground, allowing trees to grow and farmers to farm, and enabling a wide spectrum of drier and moister habitats to develop. In addition, the river does not just go about the business of drainage within a landscape, but plays an active, if unconscious, role in shaping the scenery. Were rivers left to do their work unimpeded by other natural forces then eventually all the hills and mountains would be smoothed down to their roots and the surface of the land would everywhere exist

After heavy rainfall runnels of water may combine to form a stream

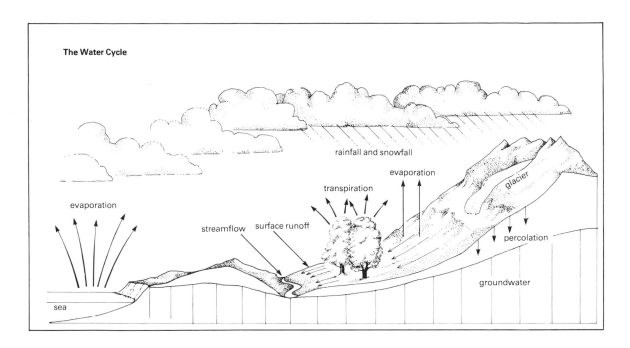

The Water Cycle

rainfall and snowfall

evaporation

transpiration

evaporation

glacier

streamflow

surface runoff

percolation

groundwater

sea

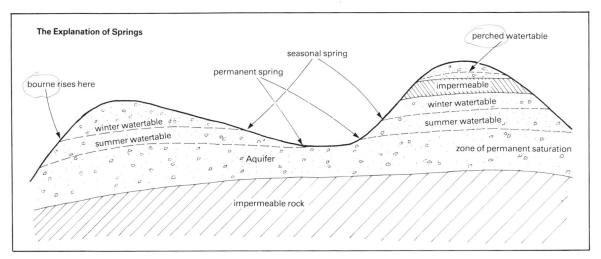

The Explanation of Springs

perched watertable

seasonal spring

permanent spring

bourne rises here

impermeable

winter watertable

winter watertable

summer watertable

summer watertable

zone of permanent saturation

Aquifer

impermeable rock

as very gentle slopes plied by slow moving rivers and streams.

River Origins

Where do rivers come from? 'From the ground', one might reply. Anyone who has followed a river up along its ever more humble headwaters to its source at a bubbling spring will know that this is true. Yet one could also reply, with equal truth, 'from the air' or, 'from the sea'. Let us

explain why this is so. Seen from space, the earth appears as a ball of brilliant blue which is blotched green and brown by land masses and streaked white with swirls of cloud. A mass of water, ejected as scalding stream when the infant earth cooled, is locked on, just above and just below the surface of the planet. And here it stays, for it cannot escape into space. The water mass is contained in a number of 'storages', but it moves perpetually between them. They are the seas

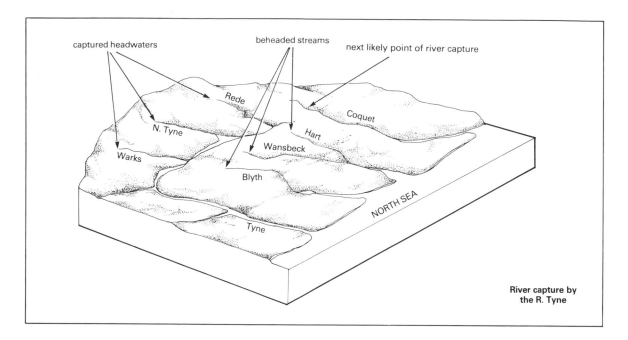

captured headwaters

beheaded streams

next likely point of river capture

Rede

Coquet

N. Tyne

Hart

Warks

Wansbeck

Blyth

NORTH SEA

Tyne

**River capture by
the R. Tyne**

A spring rising near the source of a chalk stream

and oceans, ice caps and glaciers, ground water, and the atmosphere. Rather as water might be pumped around a system of tanks, so it circulates between the storages. Some of the transfer is accomplished by evaporation, some by sleet, rain and snow, and some by streams and rivers.

Any molecule of water which is carried by a river into the ocean will eventually be evaporated up into the atmosphere and, for it, the cycle of transfer begins again. We might speculate that if environmental conditions were stable for long enough then ultimately every water molecule would have journeyed down every river and stream on earth! At any moment in time, over ninety-seven per cent of the water mass is contained in the oceans, while only one per cent is involved in the atmospheric stage of the 'hydrological cycle'. For any water droplet which is circulating in the atmosphere and about to return to earth as rain, hail or sleet (precipitation), the chances of participating further in our story by falling on land rather than dropping back to the ocean are slightly worse than one in four. The oceans cover about seventy per cent of the earth's surface, but they receive a rather higher proportion of precipitation.

A variety of different destinies may await the droplet of water which falls upon the land. It might swiftly return to the atmosphere by evaporation within seconds or minutes of its arrival. It could sink into the earth, be hijacked by the roots of a plant and soon re-enter the atmosphere via the leaves as a participant in the process of 'transpiration'. But its visit to the land could also be longer. It might percolate down

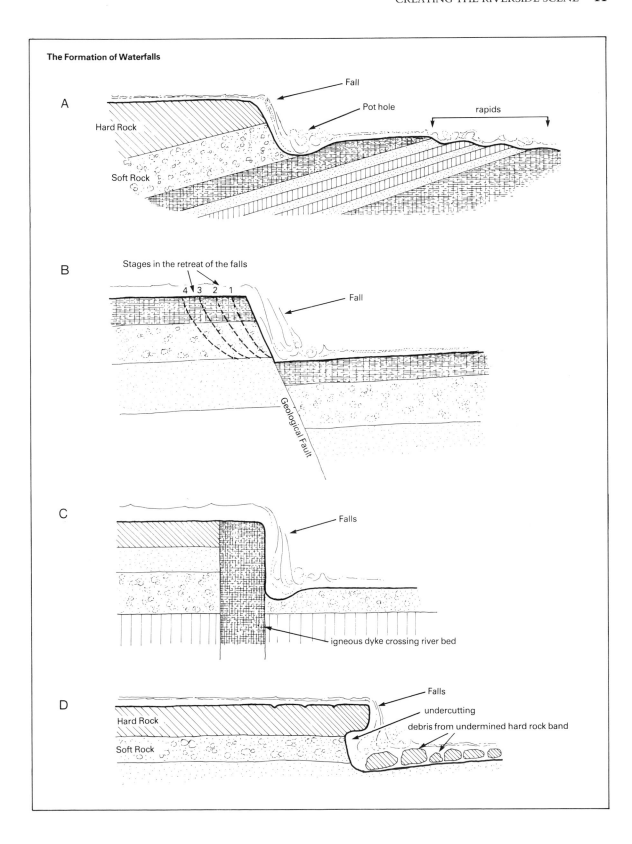

The Formation of Waterfalls

A

Fall

Pot hole

rapids

Hard Rock

Soft Rock

B

Stages in the retreat of the falls

4 3 2 1

Fall

Geological Fault

C

Falls

igneous dyke crossing river bed

D

Falls

undercutting

debris from undermined hard rock band

Hard Rock

Soft Rock

In times of flood, water pouring into the Sligachan river from the Cuillins on Skye will have the power to roll these boulders along the river-bed

through the soil and rocks to form part of a body of ground water. Most of this ground water originated in the atmosphere, some has arrived less directly, as a result of seepages from the beds of rivers, lakes and streams, while a tiny proportion is of terrestrial origin, some being released when molten rock or 'magma' cools and crystallizes. After a short or possibly a very long stay in subterranean places, ground water will 'leak' through the surface in the form of springs. The final possibility is that our droplet will become rain-wash, trickles of water flowing across rock surfaces, combining with other trickles to form runnels and rivulets which themselves merge to form a stream. All being well, the next stage is the river.

One cannot calculate the probabilities governing the different destinations awaiting falling water droplets, for the conditions on the ground beneath vary enormously from place to place and from time to time. Hard rain tends to produce a heavy 'run-off' and sustained bursts of torrential rain may be followed by flooding as the run-off quickly enters the river system. Much longer periods of steady drizzle produce less dramatic effects, since more of the rainfall seeps gently into the earth to join the ground water mass. The rocks also have an important influence; some are permeable and packed with fissures to allow the through passage of some water, and some are porous and can admit water into the tiny pore spaces. Other rocks are much less able to absorb water or assist its passage, so that ramblers caught in a cloudburst when in the tough slate and volcanic rock country of the Lake District, the gritstone crags of the Yorkshire Dales or the Cairngorms granite country will have seen the sheets of run-off sliding across the bare rock faces *en route* to a stream or burn.

These little watercourses become noticeably more animated and vociferous as the rain cascades down.

Plants too have a crucial effect, for not only do they help to drain the ground through transpiration, but also their roots and stems assist the percolation, while the protective surface mat of vegetation protects the soil beneath and disrupts the sheets of surface run-off. Other important factors include the steepness of the slopes, the level of humidity and evaporation and the degree to which the soil is dry, and able to absorb moisture, or saturated.

Already it is plain that the question 'Where do rivers come from?' is hedged around with many 'ifs' and 'buts'. Some rivers originate mainly from surface run-off collected by the branching networks of tributary streams; some derive from ground water which flows down from springs, while most derive some of their water from one source, and some from another.

Springs

Many of our most attractive little rivers and streams can be traced back to a series of springs. There is something rather magical about a spring – and in ancient times people had no doubts that magic provided the only explanation. When Christianity began to strengthen its footholds in Britain during the Dark Ages, the new religion did not spurn the opportunity to commandeer sites which were sacred to the older beliefs. As a result, many of the legends associating Christian saints with holy wells and springs have very clear pagan undertones. Death, the shedding of blood and the miraculous renewal of life which feature in the 'Christian' legends all plainly echo the older cults of fertility, magic and sacrifice. The weird blending of pagan and Christian ritual is found at St Patrick's Well or Chibbyr Pherick on the Isle of Man. Maids about to be married would fill their mouths at the well and then walk three times around the nearby standing stone, swallow the child-giving water and say, in Manx, 'In the name of the Father and of the Son and of the Holy Ghost'! Since the advent of geological science we can explain the origins of springs without recourse to magic.

One of the most attractive holy wells is formed by the spring which emerges from the limestone flanks of the Ouse valley and trickles from the churchyard wall which embraces the interesting Saxon church at Stevington in Bedfordshire. In the Middle Ages the spring was a place of pilgrimage for sufferers with eye ailments. The well and adjacent path should be visited with care for the little wetland site is preserved because of its fine colony of butterbur, a striking plant with deep pink flowers which blooms in the spring.

The upper cascade at Scale Force, above Buttermere and Crummock Water

Stevington Holy Well, Bedfordshire. The village occupies a crossroads site 5 miles NW of Bedford and is reached by side roads from the A428 and A6. The Saxon church and Holy Well are reached by the lane which runs down from the village towards the Ouse.

Water percolates downwards through pores or fissures in the subterranean rock layers until it encounters an impermeable layer. It will then stand above these beds of rock, held as ground water in the strata of an overlying water-holding rock or 'aquifer', with the upper limit of the saturated zone, the 'water-table', rising or falling according to the wetness or dryness of the weather. Very gradually, water will seep through the aquifer in a horizontal direction, perhaps only moving a hundred yards in a year. Eventually seeping water will reach a place where the aquifer ends at a steep slope or is cut by a deeply incised valley, and here the water will emerge at the surface in a chain of springs. Chalk is an extremely porous rock, and in Britain it often stands above beds of impermeable clay, and long chains of springs will be found where the geological junction meets the surface. This is particularly striking on the margins of the Lincolnshire Wolds and Lincolnshire Heights, with two parallel lines of villages set like beads on strings at the scarp face and dip-slope sides of the chalk hills, each little settlement exploiting a spring line site at the junction between the sticky clay and the parched chalk. Similarly, springs will commonly be seen where the base of a mass of limestone, a rock which is riddled by joints and cracks, meets a platform of impermeable rock.

Sometimes the downward percolation of water towards the ground water zone is delayed by an encounter with a band of rock which restricts the passage of water. Here a 'perched water-table' will form, with its own line of springs issuing from slopes standing well above the main water-table and spring line. This is well demonstrated in the Cotswolds, where an upper sequence of springs marks the place where the limestone is cut by a narrow bed of impermeable Fuller's Earth.

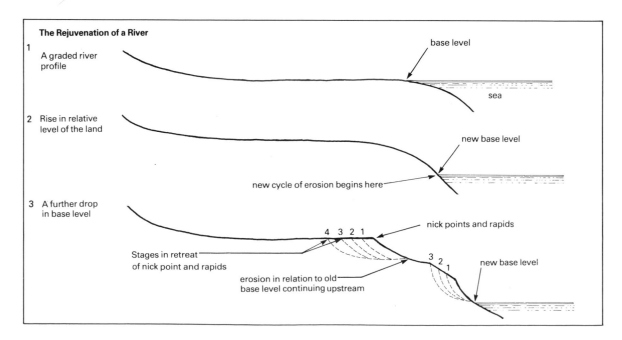

Glaciation has greatly enlarged the Glen Nevis valley at the foot of Ben Nevis

As we have said, the level of the water-table varies according to the amount of water percolating down from the surface, so that the saturated level will stand higher in winter than in summer. Fluctuations in the height of the ground water zone means that hillside springs will rise at lower levels or disappear completely in the summer.

Such intermittent springs are known by a number of charming vernacular names. The most widely used, originating in the southern downlands, is 'bournes', though in Kent the springs are known as 'nailbournes',

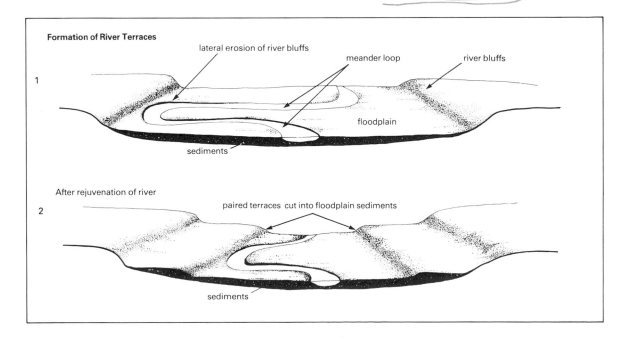

Formation of River Terraces

1

lateral erosion of river bluffs

meander loop

river bluffs

floodplain

sediments

After rejuvenation of river

2

paired terraces cut into floodplain sediments

sediments

in Hampshire, as 'lavants', in Wiltshire as 'winterbournes', while in the north of England they are sometimes called 'gypsies'.

The Headwaters

Some streams and small rivers have their origins in bournes and in lowland springs which are continuously fed by the permanent water-table. Other rivers can be traced back to streams fed by springs and by run-off which have their sources in mountain country. In such cases, the early stages of the river's course may be called the 'torrent tract'. Mountainous areas intercept the clouds rolling in from the Atlantic and receive generous quantities of rain and snow. In its earliest infancy the stream will lack the volume of water necessary to accomplish much erosion or 'corrasion' of its bed. Its waters are clear, for as yet there are few particles of sand and silt which can be used to 'sand-blast' the rocks of the channel. But quite soon, as it collects rain-wash from the bare rock faces, seepage from the sodden peat beds of the plateaux and spring water, the stream gains in muscle power. Sand particles swept along by the thundering torrents, boulders sent rumbling along the stream-bed in times of spate, and pockets of air compressed in tiny crevices by the charging waters all help to erode the channel.

Dazzling cascades form where the streams hurtle over steep rock faces and where tributary streams plunge over the sides of deep, ice-gouged valleys. Many of the finest mountain cascades in Britain can be explored in the Lake District. Here the run-off from the steep, hard rocks is very swift, so that visitors will see waterfalls in their most impressive and thunderous states

The glaciated Langdale valley in the Lake District is much too large for the modest stream it now contains

Tidal meanders on the Usk, near Newport

The beautiful cascade of Aira Force

The upper falls at Aysgarth, seen when the Ure is at its high winter level

Sourmilk Gill hurtling down the flanks of Base Brown Fell in the Lake District

Janet's Foss, a National Trust property near Malham. Sheep were once washed in the pool at the foot of the falls

Meanders in the Don near Kintore, Grampian, as seen from the air

Some Spectacular Cascades in the Lake District

All of these cascades are seen at their most impressive following heavy rain or the spring snow-melt.

Aira Force (NT), NY 400206. Reached from the A592 about 10 miles SW of Penrith. Two bridges which are reached from the car park via the wooded ravine provide fine vantage points. The main fall tumbles about 70 feet into a gorge and lesser falls are found both upstream and downstream.

Birks Bridge Rapids, NY 234994. The Duddon rushes through a narrow gorge by a little pack-horse bridge. Follow the Wrynose Pass road S from Elterwater, turning off to Seathwaite. A Forestry Commission car park lies close to the rapids.

Colwith Force (NT), NY 327031. 2 miles S of Elterwater. The Brathay flows in a series of cascades and then plunges down two parallel falls; from Elterwater the river plunges over *Skelwith Falls*, NY 341034. The approaches to Colwith Force are rather hazardous.

Dungeon Ghyll, NY 289066. A rock bridge spans a dark gorge which encloses the falls. Car parks are available on the B5343, with paths leading up to the falls. A little to the N another path leads to the falls of *Stickle Gill*, NY 174995.

Launchy Ghyll Falls, NY 306155. Two converging falls culminate in a series of lesser cascades. The falls are in woodland and can be approached from a forest track beside Thirlmere.

Levers Waterfalls, SC 283987. Church Beck leaves Levers Water and proceeds by a series of cascades towards Coniston Water. The falls can be reached via a long path from Coniston.

Lodore Falls, NY 265188. The Watendlath Beck flows down from Watendlath Tarn to Derwentwater, tumbling in the last portion of its descent. There is a coin-operated turnstile behind the Lodore Swiss Hotel which is on the B5289 lakeside road. A footpath leads through Lodore Wood to the higher falls.

Scale Force, NY 151171. A very long cascade which plunges into a dark gorge. About 1 mile from Buttermere village and reached by a path; a footbridge just downstream provides a good vantage point.

Stanley Force, SC 174995. The falls, to the S of Boot, can be reached by the path to Stanley Ghyll Beck. The beck occupies a ravine cut along a geological fault. The lower falls are more accessible than those upstream and can be viewed from a bridge.

Taylorgill Force, NY 229109. The falls can be reached via the road in Borrowdale which leads to the foot of the Sty Head Pass track; the falls are seen as a splash of white on the slopes above and the ascent is quite steep. On the slopes above Seathwaite Farm there is a smaller cascade on *Sour Milk Gill*, NY 232122.

shortly after a heavy outburst of rain or during the spring snow-melt. Two particularly fine examples are the Aira Force, a National Trust property furnished with a footbridge allowing the visitor to watch in safety the 70 foot plunge of the Aira Beck as it drops down two cascades, and Scale Force, a deep gorge with a 170 foot cataract. The locations of these 'forces' and those of other notable examples in Lakeland are given in the gazetteer.

Becoming quite potent agents of erosion, the mountain streams cut down into the bedrock to carve steep-sided gorges. As well as being able to cut downwards into the hillside or valley floor, the upland stream is gradually able to accomplish headward erosion or 'spring sapping'. This allows the stream to extend its course upslope. Occasionally, and only after a very prolonged period of spring sapping, the source may wear its way back through the original watershed to penetrate the valley beyond. It will then capture the streams flowing in this valley and so enhance its powers of valley erosion. The most famous

British example of this process of 'river capture' is found in Northumberland. Here the Tyne has captured the headwaters of the 'beheaded' rivers, the Hart, Wansbeck and Blyth, and the hijacked headwaters are now represented by the Tyne's tributaries, the Rede, North Tyne and Warks. Similarly, in the Weald near Farnham, a tributary of the Wey has captured a headstream of the Blackwater, while the Wey itself has been a victim of the piracy of its headwaters by those of the Arun, a vigorous stream system which will eventually capture more Wey tributaries.

Glaciation

The current geological period is known as the 'Pleistocene'. Thus far in the period there have been four major ice ages, while the traces of smaller glacial readvances which occurred at the end of the last glaciation are still detectable in the more northerly and mountainous parts of Britain. Contrary to popular belief, glaciation does not create entirely new scenery, but it can greatly modify the landscape. In the uplands evidence of severe glacial erosion predominates, while in the lowlands the deposition of eroded material, often in the form of great 'till plains', smeared with layers of till or boulder clay, is evident. As the climate chilled at the onset of each glaciation so increasing thicknesses of snow accumulated, causing lobes of ice to advance from the high level ice reservoirs, downwards along the existing river valleys.

Eventually, the great rivers of ice filled the valleys and the slowly advancing glaciers caused staggering feats of erosion. The alternation of freezing and thawing shattered the exposed rocks on the slopes overlooking the glaciers and the frost-shattered debris clattered down into the ice, arming it with abrasive materials which could grind and polish the ice-enveloped rocks of the valley sides and floor. Meanwhile, the ice was welded to such rocks, and as the pressures

Kilnsey Crag, North Yorkshire, SD 972678. This is a striking landmark of Wharfedale, hollowed at the bottom where undercut by glaciers. The crag lies beside the B6160 about 4 miles NW of Grassington and just N of Kilnsey hamlet.

from the ice accumulation areas above forced the glaciers forward, so fragments of rock were plucked away. The glaciers were less flexible and accommodating than the rivers whose valleys they had commandeered. Consequently the twists and turns of the valleys were often eliminated as the glaciers gradually ploughed through the interlocking valley-side spurs and transformed the V-shaped valley cross-sections into great U-shaped gouges. One of the most spectacular landforms produced by this straightening and broadening of river valleys is Kilnsey Crag, which looms above the B6160 in upper Wharfedale near the junction of the Wharfe and Skirfare, the cliff face of the crag cut, steepened and polished by the great glaciers which once filled the river valley.

When the ice retreated, the lowland rivers were sometimes obliged to form new courses, picking a path across the chaotic plains of undulating debris which had been ground from the uplands and dumped in the plains and lower valleys. The almost lifeless landscape must have looked like some vast and badly managed building site. In the uplands the rivers adjusted themselves to a landscape reworked by the ice. Accumulations of soil had been stripped away, while the ice had attacked the softer strata of the valley floors, creating great over-deepened hollows which now became lakes. Sometimes the places where advancing ice had stabilized were marked by hummocks of debris or 'moraines' and the rivers were ponded-back until they could cut a course through the barrier. The erosive powers of the glaciers were strongest in the larger valleys with the greatest accumulations of ice. Tributary valleys which had accommodated smaller glaciers consequently experienced less gouging, and as a result they now existed as 'hanging valleys', their courses suspended high above the floors of the much-deepened main valleys.

The original ice accumulation hollow at the head of each glacier was severely eroded, being carved into a basin-like 'corrie' or 'cwm'. Often such corries now exist as lakes or tarns and may act as natural reservoirs at the heads of mountain streams. One of the most spectacular of the hundreds of examples in Britain is Red Tarn in the Lake District. To those who have climbed to the summit of Helvellyn it looks like a puddle at the bottom of a gigantic rock basin, while those who make the long plod up to Red Tarn's shore from Patterdale see Striding Edge looming like a gigantic cock's comb above the still, dark waters.

Frequently the modest streams which reoccupied the glacial troughs inherited a valley which was several sizes too large; Bishopdale in the Yorkshire Dales is a good example. This discordance between the humble tenant and the mightly predecessor can be seen in many mountainous places, and beautifully imposing examples include the magnificent Llanberis Pass in Snowdonia and the Langdale valley in the Lake District.

The creation of post-glacial lakes by the over-deepening of pre-existing river valleys is, of course, most wonderfully demonstrated in the Lake District. Ever since the glaciers were converted into torrents of melt-water the rivers have 'sought' to undo the work of the ice by filling the lakes with sediment. Although deglaciation occurred quite recently – a mere 10–12,000 years ago – this work is now quite well advanced. Consequently several of the lakes are fringed by level expanses of green farmland which, following the enclosure of the old hamlet open fields, are now patterned by a delightful tracery of grey stone walls. Perhaps the best place from which to view this kind of scenery is the mountain slope overlooking Rosthwaite: the rock bastion and Iron Age hillfort of Castle Crag provide excellent vantage points. Here the little Derwent has gradually deposited sediments which are pushing the shores of Derwentwater northwards, and which may one day completely fill the lake. A splendid view, including the present outlet of the river in the lake and mountain panorama, is the 'surprise view', enjoyed by visitors standing on the edge of the precipitous cliff above Ashness Bridge.

The ferocity of glaciation varied greatly from place to place. The glaciers were most potent in lofty mountain areas where altitude and the steepness of slopes were combined with exposure to the snow-laden Atlantic cyclones. Along the seaboard of the Western Highlands of Scotland and in the Cuillins of Skye the glaciers plummeted from the high peaks into a sea – slightly shrunken by the locking of water on land as ice – where the 'calving' of the glacial snout as the ice met the waves accelerated the rate of ice advance. As the vigorous glaciers met the old shoreline deep trenches were carved and, following the post-glacial rise in sea-level, these now exist as fjords or sea lochs. Consequently, many Highland rivers, like the Fyne, Etive, Torridon or Broom, have a brief but eventful existence before terminating in the salty waters of a tidal sea loch. In the Grampians and Cairngorms there was less exposure to snowstorms and the gradient between the summits and the North Sea was much shallower. Consequently the effects of glaciation are far less dramatic and the depositional features predominate. One truly dramatic glacial feature is the Burn o' Vat near Dinnet, a great cauldron and gorge gouged and drilled into the landscape by ice and melt-water rather than by the modest stream which now descends from the Culblean Hills.

Thornton Force, part of the beautiful Ingleton series of falls

Waterfalls and Rapids

Waterfalls can occur at any point in the course of a river or stream, although they are most commonly associated with the more youthful and exuberant upper reaches. A cataract can be seen where a stream plunges down a steep cliff face or tumbles from a hanging valley, not yet having had the time or the muscle power to carve a gentler course along the bed of a gorge. In other cases, waterfalls are created by geological conditions. Where a vertical band of tougher rock – perhaps an igneous 'dyke' intruded into a fissure in an older rock mass – intersects a river course, then a waterfall may form where the river plunges over the edge of the hard rock and vents its erosive fury on the adjacent softer rock. Alternatively, chains of cascades and rapids may form where a series of rock beds of varying toughness outcrop along a river-bed. In such conditions the river will erode the less-resistant rocks more rapidly than their tougher neighbours, carving a sequence of falls, potholes and steps. Finally, a waterfall will form when the river flows across a horizontal bed of resistant rock which stands upon softer strata. At some point the hard bed will terminate, and here a waterfall will develop as the river exploits the softer rock. As it does so, the pebbles and boulders which are swirled around in the gushing waters will pound and grind the softer beds, undermining the lip of the hard rock above until an undercut block topples. As the process of undermining continues, so the waterfall very gradually migrates upstream, and as it migrates a steep gorge forms, marking the slow recession of the falls – as can be seen at Scale Force and many other lovely waterfall sites.

Perhaps the finest place to explore the story of the waterfall is at Ingleton Falls near the spectacular mountain of Ingleborough in the Pennines. This is something of a Mecca for geologists, displaying incredibly old green Pre-Cambrian slates and sandstones which make their 330 million-year-old Carboniferous conglomerate and limestone neighbours seem like youngsters, as well as a dyke of pinkish rock and blue-black slate. The rapid alternation of rock bands of different degrees of resistance produces a fascinating series of beautiful cascades, with the tougher sandstones forming the waterfall structures. The cataracts increase in beauty and drama as one follows the path upstream, culminating in the splendours of Thornton Force, the only fall on limestone. The locality includes plenty to fascinate the visitor, the limestone extravaganzas of the Ingleborough area such as the White Scar caves, Gaping Ghyll pothole and Alum Pot, with the breathtaking cascade of Alum Pot Beck.

Also in the Yorkshire Dales, there are a number of fine falls on the Ure. Best known is Aysgarth Force, where the waters plunge over three great limestone steps. The surrounding banks, though now somewhat the victims of trampling, display a cream and purple mosaic of primroses and violets in the spring, while the invading mimulus or monkey-flower blooms in waterside niches during the summer. Hearne Beck, a tributary of the Ure, plunges over a limestone cliff at Hardraw Force, and here it is possible, though inadvisable, to walk behind the sheets of falling water. Less imposing, but in gentle wooded country which contrasts well with the barren majesty of nearby Gordale Scar, is Janet's Foss. This is a National Trust property near Malham in Yorkshire, a place where sheep were dipped in the pool at the foot of the falls. Further north, High Force on the Co Durham and Yorkshire boundary is a spectacular cataract which slithers and plunges down a dark cliff of dolerite. Also on the Tees system and not so far away, though less accessible, is Caldron Snout. Here the river descends a longer dolerite staircase; the pool known as the Weel which fed the falls has now been engulfed by the controversial Cow Green Reservoir. The northern uplands enjoy a virtual monopoly of impressive waterfalls in England; elsewhere in the country there is little of great note, though mention can be made of the Lydford Cascade or 'Woman in White', where a tributary stream drops from Dartmoor to join the Lyd.

In the Scottish Highlands, where waterfalls are sometimes detectable on maps from their Celtic name *eas* there are many fine examples. These include Britain's highest falls with a drop of about 785 feet at Eas Coul Aulin at the head of Glencoul in the old county of Sutherland. More accessible are the Falls of Foyers south of Loch Ness and the Falls of Clyde near Lanark, both adversely affected, from the tourist point of view, by the control of their waters by hydroelectric schemes. The

The dramatic cascade at Hardraw Force

National Trust for Scotland owns the area where wild goats roam around the Grey Mare's Tail Falls near Moffat in Dumfries. Here the Tail Burn plunges into Moffat Water. Wales too has many fine cataracts tucked away amongst spectacular mountains. The more accessible examples include the majestic Pistyll Rhaeadr at the head of the Afon Rhaeadr, reached by a track from Llanrhaeadr-ym-Mochnant to Tan-y-pistyll; the Swallow Falls, above the junction of the Conwy and Llugwy near Capel Curig, and, in the south, the Henryd Falls on the Nant Llech, a tributary of the Tawe. Space does not permit a mention of all the fine falls in Britain, but room must be found to note the lovely Glencar Falls by the side of Glencar Lough near Sligo in the Republic of Ireland – a strong contender for the title of the most beautiful cascade in the British Isles.

A Selection of Britain's Finest Waterfalls

The Ingleton Falls, North Yorkshire, SD 695750. These lie on the N side of Ingleton and the entrance to the car park is off the B6255; there is a small charge. As one walks upstream, Pecca Fall comes into view, and further upstream are the still more spectacular falls of Thornton Force.

Aysgarth Force, North Yorkshire, SE 022888. The falls lie to the E of Aysgarth village, where a car park is available, and are reached by footpaths which offer superb views of the upper and lower falls.

Hardraw Force, North Yorkshire, SD 870917. Here the Hearne Beck, a tributary of the Ure, plunges from the top of a curving cliff. Access is gained through the hall of the roadside public house and the force is well signposted. In cold weather the immediate approaches to the falls become coated in ice and great care should be taken.

Janet's Foss (NT), Cumbria, SD 913634. On the opposite side of the minor road from Malham to Gordale Scar. A path follows the beck which falls 15 feet into a round pool.

High Force, Co Durham, NY 880284. The falls are about 5 miles NW of Middleton-in-Teesdale and reached by a wooded path with its entrance on the B6277 opposite the High Force Hotel. The Tees drops 70 feet over the dark cliff of the Great Whin Sill.

Caldron Snout, Co Durham, NY 814289. This is to the NW of High Force and can be reached by turning off the B6277 at Langdon Beck. After driving for 2 miles the track is followed on foot for a further 1 mile. The river falls down a dolerite staircase for 200 feet.

Lydford Cascade (NT), Devon, SX 500835. The falls lie 7 miles N of Tavistock, W of the A386 and are best approached from the car park beside the bridge in the S of Lydford.

Eas Coul Aulin, Highland, NC 273278. Accessible only by boat and footpath one ascends the steep slopes on the S side of Loch Glencoul and above the sheer cliffs beyond the head of the Loch to encounter Britain's highest waterfall.

Falls of Foyers, Highland. The falls lie just to the S of Foyers, which is E of Loch Ness on the B862. Foyers was the setting for the first commercial hydroelectric scheme in Britain, which began operation in 1896.

The Falls of Clyde, Strathclyde. These falls lie just to the S of Lanark.

Grey Mare's Tail Falls (NTS), Dumfries and Galloway. Three Scottish falls share this name. These lie E of the A708, 8 miles NE of Moffat, where the Tail Burn plunges into Moffat Water.

Rapids of Glen Nevis, Strathclyde. A minor road from Fort William runs SE into Glen Nevis and is used by tourists seeking to climb Ben Nevis. The rapids lie in the gorge beside the final stages of the road.

Falls of Bracklin, Central. The falls are to the NW of Stirling, E of the A84 near Callander. A footpath leads to the footbridge crossing the Keltie Burn.

Pistyll Rhaeadr, Clwyd, SJ 074295. The falls are 4 miles NW of Llanrhaeadr-ym-Mochnant at Tan-y-pistyll.

Swallow Falls, Gwynedd, SH 765578. These falls lie 3 miles WNW of Betws-y-coed on the A5. They can be approached by a footpath following the N bank of the river.

Henryd Falls (NT), Powys. The falls are 11 miles NE of Neath, just N of Coelbren. A minor road leads from the A4067 at Pen-y-cae to the deep wooded ravine in the Graigllech Woods and to the 90 foot falls.

Rapids can occur where a band of tougher rock outcrops in a river-bed, but to understand properly the formation of rapids we must delve a little more deeply into the sciences of fluvial geomorphology and hydrology. Rivers are so vibrant and personable that it is easy to make the mistake of regarding them as living organisms. If rivers were alive then one could say that the work of a river is undertaken with the goal of carving a completely smooth course, free of any falls or rapids and perfectly graded from source to outlet. The long cross-section or 'long profile' of the river would, ideally, be smoothed and lowered so as to leave just sufficient gradient and velocity to allow it to move the burden of silt in its waters. Each river erodes its bed down according to the level of its outlet. This is known as 'base level' and is normally represented by the sea, although lakes and waterfalls along the river's course can provide temporary base levels.

Were conditions in the natural world to remain stable for long enough, then eventually each river would accomplish the task of carving a perfectly graded course, with the long profile of each river existing as a smooth parabolic curve. The world, however, is not a stable place. In time mountains surge upwards and faulting raises and lowers great blocks of country, so that the relative positions of land and sea are constantly changing. If there is a relative rise in sea-level then the lower course of a river will be invaded and a new base level is established. If there is a relative rise of the land, then river erosion must adjust to the new base level and steeper gradients. The cutting of a new river profile begins at base level and, very gradually, river erosion causes the new profile to retreat upstream. The junction between the old and the new profile is marked by a 'nick point', with rapids, while above these rapids the river continues its grading work in relation to the old river profile.

Often the effects of several changes in base level will be evident along a river course, each change marked by a set of rapids, and each set of rapids migrating slowly upstream towards the river's source. And so the erosive work of the river might be compared to that of a carpenter who is seeking to shape a smooth curve, but finds that his timber is periodically moving in the vice, and each time he is obliged to begin carving the curve again.

Britain contains hundreds of splendid sets of rapids, many of them occurring near the junction of the torrent tract and middle or valley stage. The Falls of Bracklin in the old county of Perthshire, where the Keltie Burn negotiates a spectacular sequence of falls, pools and rapids, are one of the most impressive examples. Another fine set of rapids can be seen by ramblers moving up to the head of Glen Nevis *en route*, hopefully, for the summit of Ben Nevis. These rapids slide across rocks which are honed and fluted into strange forms by boulders which crash and thunder in the pounding waters.

The Valley Stage

As the river passes from its torrent tract into its valley stage it is enhanced by the union with its various tributaries. The increasing volume of water raises the potential for erosion, but meanwhile, as the gradient decreases, so the main assault of the erosive forces switches from the corrasion of the river-bed to attacks on the sides of the channel. By this stage in the river's course, however, most of the irregularities in the bed which would cause falls have been removed, while the velocity of water flowing in the channel gradually becomes insufficient to ensure that the bed is swept clear of sand and silt. Slowly, down-cutting is superseded by lateral erosion, and as the river swings from side to side across the valley floor, so it undercuts the banks and hillsides. What was once a narrow V-shaped valley is gradually moulded into a valley flat which is flanked by lines of steep river bluffs as the valley spurs are worn back and the valley flat widens.

When its tributaries are gorged with floodwater the river will overflow its channel, and as the flood abates, so river silts are deposited across the valley flat or floodplain. But if there is a drop in base level, caused as we have described by a relative rise in the land or drop in the sea, then the erosive power of the river will

be increased. Thus invigorated or 'rejuvenated', the river will begin to cut down into its old floodplain, swinging from side to side and undercutting its banks, but also sweeping its bed clean of sediment and beginning to attack the rocks beneath again. The junction between the old floodplain level and the new, lower, floodplain is marked by a step, usually just a few feet in height, and the valley floor now includes a set of 'river terraces'. A new cycle of rejuvenation will repeat the process, and eventually the valley bottom may contain many sets of terraces. Sometimes each set is distinct and clearly recognizable, with the sections of old floodplain stacked at different levels. In other cases, as the river weaves across its floodplain some of the old terraces are destroyed, so that in places the terraces form shallow staircases leading down to the

Meanders in the Herefordshire Wye, as seen from Goodrich castle

river from the bluff line, while in other places they have been obliterated and the river may be seen undercutting the valley sides and pushing back the bluff line.

These then are the components of the 'classic' valley scene, admired, if not always understood, by all country lovers in countless lovely places. The river winds across a broad floodplain which is bounded by bluffs and whose surface consists of a series of long, low terrace steps. Such valley landscapes are common in the countryside, but they are also a feature of the London townscape, even though the buildings obscure the pattern in the scene. Kew and Barnes stand upon the youngest and lowest river terrace; Crayford occupies the next step in the terrace staircase; Swanscombe the next, while Dartford Heath stands upon the uppermost and oldest terrace. Across the river to the north, Ilford stands on a terrace which can be paired with the one to the south supporting Crayford. Elsewhere, uncluttered sets of river terraces can be seen further upstream in the valley of the Thames, in sections of the valleys of the Severn and Wye, and in scores of other places.

The Plain Tract or Lower Course

In its final stages the typical river flows across a broad lowland plain before its waters make their inevitable rendezvous with the sea. Here the gradient of the river has become so slight that it lacks the power to erode its bed, and all the energy of the slow moving waters is absorbed in the task of supporting the burden of fine silt washed down from the upper reaches. As the currents slow, the river is unable to hold the mud which is suspended in its clouded waters and it is deposited on the river-bed. Although it has lost the power to deepen its bed, the river is still capable of lateral erosion, while the lack of a distinct gradient causes it to become rather aimless and rambling. Great meander loops form as it weaves drunkenly across its vast floodplain. As innumerable schoolchildren are obliged to learn, often the meanders become so exaggerated that the progressive narrowing of the neck of the meander culminates in the union of the two sections of river at opposite sides of the meander neck, so that the remainder of the meander is left stranded and stagnant as a cut-off or 'ox-bow' lake. Yet the rejuvenation of the river is always a distinct possibility, and where it occurs then the envigorated river, its power now enhanced by the steeper gradient, is able to recommence the erosion of its bed, so that the meanders become incised into the floodplain.

River meanders can be impressive, even

graceful features when seen from the air, but because they usually exist within broad, low plains, good terrestrial vantage points are uncommon. Even so, meanders are very common features of the lowland scene, particularly well marked in the lower courses of such rivers as the Tamar, Tees and Forth. A fine view of tidal meanders on the Usk is gained from the Caerleon to Newbridge-on-Usk byroad near Newport, while a 'text-book' series of meanders can be seen on the Cuckmere between Lewes and Eastbourne.

As the senile river weaves and staggers across its floodplain, mud is deposited upon the bed, often actually raising it above the level of the surrounding countryside. In flood the river will burst its banks and inundate the surrounding expanses of flat land. At such times, however, silt is deposited on the adjacent riverside areas, gradually building up broad, low natural embankments or 'levees' which help to hold the river to a disciplined course. Many of the levees flanking British rivers are artificial formations, built from mud dredged from the silt-choked bed and dumped upon the banks to improve navigation or flow and reduce the threat of flooding.

Where the river terminates at an estuary one may either encounter a fascinating world inhabited by plants and animals especially adapted for life at the junction of two natural worlds, or else a complex of foul creeks and barren mudflats rendered lifeless by the works of man. Occasionally, however, the union with the sea is postponed. A superb example is that of the Alde, south of Aldeburgh in Suffolk. Here 'longshore drift' accomplished by sea currents which have swept sand and shingle southwards along the East Anglian shore have created the ten-mile-long spit of Orford Ness. The Alde (or Ore) is diverted to follow the shoreward edge of the spit for its entire length before releasing its waters into Hollesley Bay.

Just a little further down the coast the scenery is quite different, for here we encounter a coastline of submergence where the old river valleys have been partly inundated by the sinking of the land level, giving rise to the finger-like inlets or 'rias' which drown the lower courses of the Deben, Orwell and Stour near Ipswich, and the Colne and Blackwater south of Colchester. Other good examples of rias can be seen along the south coasts of Devon and Cornwall and the coast of Hampshire, where the rias provide sheltered anchorages for yachtsmen, and in the south west of Ireland.

How Old is a River?

Often the answer to this question is, 'far older than one can imagine'. Once a river

Goring Gap, where the Thames has sliced through the chalk scarp of the Chilterns

has established its niche in the landscape it will endure for millions of years. Geological traumas, such as episodes of mountain building or the filling of a sedimentary basin, may succeed in disrupting whole river systems, but sometimes the geological transformations may occur at such a slow rate that rivers, rejuvenated by each rise in the land, are able to erode their beds and settings in a way that keeps pace with the great earth movements.

The essential topography of the Lake District exemplifies the durability of a basic pattern of drainage. About 300-260 million years ago the marine sediments of this area were cast upwards to form a massive mountain dome. Immediately the rivers which were born established a radial drainage pattern, draining outwards from the crown of the dome-like spokes from the hub of a wheel. Long ago they stripped away the Carboniferous Limestone which blanketed the dome and then superimposed their courses on the hard old slate and volcanic rocks beneath. Even today the basic radial pattern of drainage is preserved by the rivers and the young glacial lakes which occupy parts of their courses.

In their youth all rivers are delicately adjusted to the existing terrain and geology, but as they strip away the uppermost layers of strata then they may superimpose their courses upon the rocks beneath – as we have seen in the case of the Lake District. When this occurs the rivers preserve their original courses, but these courses bear no relation to the characteristics of the rocks exposed, with rivers slicing right across exposures of tough rock. Sections of the Wye in the former county of Monmouth are superimposed upon the geological roots of scenery. The river has cut a course down through the rocks on the western margins of the Forest of Dean, and it is said that fossil meander loops can be recognized in places which stand more than 325 feet above the present course of the river. Another good example is that of the Thames at Goring Gap, where the river sliced through the pre-existing strata like a geological rip-saw and can be seen to cut right across an impressive chalk escarpment. Here the river flows through a wooded gorge which has divided the chalk ranges of the Chilterns and the Berkshire Downs.

While a river may hold to its course for millions of years, geological events may exert an enormous effect. Occasionally the changes may happen very quickly. It was only around 12,000 years ago that the Thames flowed across post-glacial marshes and mudflats which filled the southern half of the North Sea basin. The Thames and Rhine combined their courses to form a mighty river in the marshes between the present Wash and the Hook of Holland, while the Humber and the Great Ouse may have united in the former lowlands somewhere about fifty miles to the east of the present port of Hull.

CHAPTER 2

Rivers of Chalk and Limestone Country

The character of a river or stream is greatly influenced by the terrain and geology across which it flows. This simple truth is most vividly demonstrated in the landscapes where the watercourses traverse – or attempt to traverse – calcareous rocks like chalk and limestone. These are composed of lime, the chalky shells or skeletons of sea creatures and fossilized coral reefs. Depending on the particular rock concerned, the landscape may be one of dry downlands with clear rivers which are vibrant with life flowing in the troughs between the escarpments, or else a more spectacular assemblage of gaping potholes, plunging gorges and cavern labyrinths. In either case, the traditional countryside in such areas was one of bright green, closely cropped sheep pastures. But in so many places modern agricultural policies have taken a heavy toll of the emerald setting.

Chalk Streams and Winterbournes

Chalk is one of the youngest and commonest rocks in England, though rare in the Celtic countries, where older rocks predominate. Sometimes it can be seen forming plateaux and escarpments, as in the Yorkshire Wolds, the Lincolnshire scarplands and the southern Downs, while sometimes its presence in the landscape is masked by a surface smear of boulder clay, as in some eastern counties. Before it was eroded away to reveal the rocks beneath or

An enormous dry valley or 'combe' in the chalk escarpment above Westbury in Wiltshire

blanketed in sediment accumulating on the lower ground, the chalk was even more extensive. Most was deposited during the Cretaceous period, around 150 million years ago. At this time the land that is now Britain was largely submerged by a sub-tropical sea, and conditions were similar to those experienced in many parts of the Caribbean today. Such seas throng with microscopic sea creatures with chalky shells and skeletons, and the sea bed receives a constant rain of minute, lifeless bodies. At the same time the activities of bacteria deposit chalk as a chemical precipitate of lime. The deep bacterial ooze accumulated in sea-floor layers, which could be many hundreds of feet thick and which were compacted by the enormous pressures from above. Subsequently, the old sea-floors were uplifted and exposed as dry land, and the young river systems commenced their work of stripping away the chalk domes.

In the pursuit of this task the rivers were aided by the fact that the almost pure calcite or calcium carbonate which composed the chalk rocks is very susceptible to solution by water. It has been estimated that while the Thames brings down about a quarter of a million tons of sediment in suspension in its waters each year, it also removes twice this amount of material – largely chalk – by solution. 'Carbonation' is a chemical process which involves the conversion of carbonates, like calcium carbonate, into far more soluble bicarbonates. Rain-water – particularly when the clouds have been charged with industrial atmospheric pollution products – takes carbon dioxide from the air and becomes a weak form of carbonic acid, while water percolating through the humus layer in the soil becomes diluted humic acid. These acidic waters react with chalk and limestone rocks, converting their calcium carbonate into the much more vulnerable calcium bicarbonate. Striking examples of the potency of this form of weathering can be seen in towns where

Jurassic limestone, a superb building and monumental stone, has been used in the construction of medieval buildings and statuary. The best qualities of the stone will 'case harden' and develop a tough outer skin when exposed to the atmosphere, but the ravages apparent in the stonework of some Oxford colleges, and the statuary on the front of Wells Cathedral prior to the recent restoration, and evident in countless other places, testify to the vulnerability of calcareous rocks when exposed to acid rain. The threat increased enormously with industrialization, so that some of the extensive restoration work done in late Victorian and Edwardian times has already decayed to the point where new restoration is needed.

Yet both chalk and limestone terrain have their own in-built defence mechanisms to counter the attacks of surface running water. Chalk is an extremely porous rock – a fact established by countless schoolchildren engaging in the timeless pursuit of dipping chalk in an inkwell. It can absorb up to half its own volume of water, so that much of the rain or snow-melt on the chalk surface will rapidly sink down to the water-table and emerge elsewhere as springs. Chalk is often heavily fissured in places where the strata have cracked under the strain of earth pressures and movements. In such places the rock becomes highly permeable as well as extremely porous. By absorbing so much of the surface water, chalk reduces the ravages of river erosion.

In typical chalk country we find that the downland scarps and plateaux are virtually devoid of surface streams, though springs issue from the places where the chalk meets less porous or permeable strata. Other springs mark the position of the water-table, and, as we have described, winterbournes, nailbournes and lavants will flow when the water-table rises in winter. Another obvious characteristic of downland is the presence of numerous dry valleys. Clearly these barren

A chalk stream in spate, the Frome above Dorchester

troughs were once cut by flowing water, and in the limestone country which we will soon encounter it is plain that rivers which once flowed across the land surface have discovered inviting fissures and adopted underground courses. Sometimes this can also be seen to happen in the chalklands. In Surrey, between Dorking and Mickleham, the aptly named Mole vanishes in the summer, taking its leave via a sequence of twenty swallow-holes which serve as portholes on the subterranean world. Could this form of disappearance be a common cause of dry valleys in the downlands, or do these valleys commemorate times when the rainfall was sufficiently heavy to allow rivers to traverse the land surface? Perhaps the best answer can be found by reference to the Ice Ages. During the peak glacial eras, ice sheets and glaciers affected all of Britain apart from the southern third of England, where a bitter 'periglacial' climate prevailed. Under such chilly regimes the soil and ground beneath were frozen to a great depth. Being sealed by the ice in the permafrost layer, the chalk would have been unable to accept the torrents of melt-water which flowed in

summer, so that surface valleys were cut across the refrigerated countryside.

In some places the chalk surface carried a thin sheet of younger but extremely tough sarsen rock. It appears that under the periglacial conditions, when in the summer thaws a narrow slithery layer stood upon the rigidly frozen permafrost below, great blocks of sarsen would glide down into the valley hollows, like stately galleons plying a sea of mud. In due course such sarsens were used in magnificent Wessex monuments like Stonehenge, Avebury or the megalithic tombs of West Kennet or Wayland's Smithy. Dry valleys with a sarsen litter which have escaped the attentions of modern farming endure in a few places. From the roadside near the Hardy monument in Dorset there is a good view of the Valley of Stones. Here the sarsen litter in the dry valley bottom is flanked by the impressive earthworks of small rectangular prehistoric fields, part of a magnificent network which extends up to the skyline –

A spectacular expanse of limestone pavement above Malham Cove

so perhaps some of the stones were not dumped by the periglacial processes described, but are the debris from ancient stone-picking activities associated with the cultivation. Beside the A4 just to the west of Marlborough the National Trust manages Piggle Dene and nearby is Lockeridge Dene, dry valleys where the scattered sarsens are known as 'Grey Wethers' because of their resemblance to flocks of grizzled sheep.

Chalkland rivers have their origins in the springs which gush forth from the lower flanks of the downland scarps and plateaux. The base-rich waters present a striking contrast to the acidic streams of most upland areas, and chalk streams are noted for their plants and wildlife. Many a southern chalk stream has become legendary in angling circles for the quality and guile of its brown trout population. There will be no shortage of anglers who are happier fishing a slender chalk stream with its dappled pools

festooned with traps for the unaccomplished cast and its fish as cautious and fastidious as any, than they would be on the most exclusive Scottish salmon river or renowned Irish lough. An unfortunate consequence of the prestige of many chalkland rivers and

Some Sarsen Streams

Valley of Stones, Dorset, SY 595873. There is a good view of the sarsens from the Hardy Monument-Littlebredy road. Remarkable networks of Celtic fields cover the surrounding slopes.

Lockeridge Dene (NT), Wiltshire, SU 145673. The sarsens lie in a small valley beside the side road from the A4 (which leads southwards from the main road a couple of miles W of Marlborough), just to the SW of Lockeridge village. *Clatford Bottom*, with a fine collection of sarsens lies immediately to the N of the A4 at SU 155690, and other sarsen streams nearby are on Fyfield Down, just W and NW of Clatford Bottom and E of Avebury, between Marlborough and Winterbourne Monkton.

streams, and of the wealth and influence of those who fish them, is the fact that scores of old riverside paths have been lost and the banks are studded with hostile signs and patrolled by bailiffs who are ever ready to evict the harmless rambler. And so it is particularly fortunate that the National Trust manages a number of riverside properties in chalk stream country. The country-lover who is remorselessly chivvied away from the lovely Test can at last set foot on the riverside at Mottisfont Abbey, upstream from Romsey, and the Trust also owns countryside properties near the estuary of the little Hamble, south east of Botley and at Waggoners' Wells near Ludshott, where a string of hammer ponds forms one of the sources of the Hampshire Wey.

In days gone by the chalkland landscape – with its bright rivers flowing with a gentle sparkle through bands of lush water-meadows and the flower-spangled downland beyond – was as lovely as anything which the mellow south could

The resurgence of a limestone stream on the flanks of Pen-y-Ghent in the Pennines

Some Riverside National Trust Properties in Hampshire

Mottisfont Abbey. The house lies 4½ mile NW of Romsey and ¾ mile W off the A3057. A 12th century Augustinian Priory, converted into a mansion after the Dissolution, with lovely parkland, trees and old roses in gardens beside the Test.

Hamble Estuary, SU 525120. The Trust manages an area of countryside along the N of the river, directly W of Curbridge village, which is on the A3051 to the E of Southampton and SE of Botley.

Waggoners' Wells. The Trust has 40 acres at Bramshott Chase, just N of Liphook, which is on the A3(T) between Petersfield and Haslemere; nearby is Ludshott Common and Waggoners' Wells, with hammer ponds from the old iron industry forming a source of the Wey.

offer. Many of the streams are still hemmed by water-meadow pasture, although the old flooding techniques which stimulated the grass into early growth are now abandoned. An attractive sequence of old water-meadows lines several Wessex chalk streams, like the Frome above Dorchester. But the lovely chalkland river setting has become a thing of memories. Recent figures from the Nature Conservancy Council show that since 1947 the nation has lost eighty per cent of its sheep-grazed calcareous pasture (as well as ninety-five per cent of its herb- and flower-rich meadows). Responding to the inviting state and EEC subsidies, agriculture has turned the old closely cropped downlands into great featureless prairies, which stand a deathly grey or wheaten yellow according to the seasons. The Marlborough Downs can be as dull as Alberta or Wyoming, while the

Typical limestone country in Littondale, with bright pasture, bright exposures of limestone in the scars and stream-bed, stone walls and diminutive streams

Floodlit calcite formations at Stump Cross Caverns

The dry limestone gorge of Cave Dale above the Peak Cavern. Castleton can be seen in the notch of the gorge

A waterfall thunders through
a hole punched in the limestone
wall above the lower falls in
Gordale. This is a dangerous
place and visitors should
proceed with great caution

The Chess near Chenies,
a delightful chalk stream
of the Chilterns

country-lover driving across the Dorset downs sees the once-stunning networks of prehistoric fields now existing only as soil marks in the dead prairie. Over much of Norfolk and Wiltshire, Kent and Sussex the story is the same, while on the Yorkshire Wolds the soil-wash from ploughing leaves the crop standing in the ever more abundant chemical fertilizers, and little else. To see downland today which resembles the scenes of our childhood or parents' times one can go to an open air museum, the Queen Elizabeth Country Park near Petersfield in Hampshire. Surviving fragments of the old chalkland flora can be enjoyed by those with the necessary membership or permissions at nature reserves like Aston Clinton clunch pits or Totternhoe Knolls in the Chilterns, while in battered Bedfordshire the National Trust has saved 286 acres of Dunstable Downs and also 136 acres at Sharpenhoe in the Chilterns, with fine views towards the Ouse.

The new farming practices in the downlands have important consequences for the chalk streams as well as for their settings. In East Anglian chalklands the use of underground water for costly crop irrigation schemes as well as the domestic demand can lower the water-table considerably. (For several years the authors lived in a cottage with a deep well which served the household when the cottage was built in the 1870s, but which is seasonally dry today.) Water seldom flows now at the 'official' source of the Thames. The exposure of the bare plough soils of the rolling downland to the winter rains encourages a rapid slope-wash from the hedgeless landscape and the consequent silting of the formerly clean-swept river-beds, while the introduction of copious quantities of fertilizer and other agricultural chemicals disrupts the river ecosystems and, according to some scientists, may cause the threat of stomach cancer from human water supplies. Because the potential rewards for ploughing the

Carboniferous Limestone plateaux of the north and west are much less, the devastation there is much more localized. Even so, the EEC report already quoted notes that since 1947 we have lost forty-five per cent of our exposures of gaunt, creviced limestone pavement, much of the destruction resulting from the unthinking demand for rockery stone.

The Dramas of Limestone Country

Limestone is another very common rock. It comes in different types and ages, most notably the Jurassic limestone of places like the Cotswolds and Northamptonshire Uplands; the older Magnesian Limestone which hems the eastern edge of the Pennines; the Carboniferous Limestone of some of the northern and western uplands, and the ancient Durness limestone, which makes a brief appearance in the far north west of Scotland. It is the Carboniferous Limestone which is associated with the most spectacular water-cut scenery. This is a silvery grey rock which also originated in the deposition of calcium carbonate on the floor of an ocean, but it is far older than the chalk, and also considerably tougher. Accumulating at rates which were probably less than a couple of feet every thousand years, this limestone layer eventually amassed to a thickness of well over a mile. Today the limestone, often more resistant than neighbouring rocks, can be seen forming upland plateaux which are gouged by gorges and edged by cliffs or 'scars' in the Yorkshire and Derbyshire Dales, the Burren of Co Clare, the Mendips, and also in more localized outcrops like the Great Orme near Llandudno or Eglwyseg Mountain near Llangollen.

Despite its toughness, the limestone is susceptible to water erosion. Just like the chalk it is prone to solution and carbonation, while various traumatic episodes of compaction, uplift, folding, faulting and the removal of the overlying rock burden have

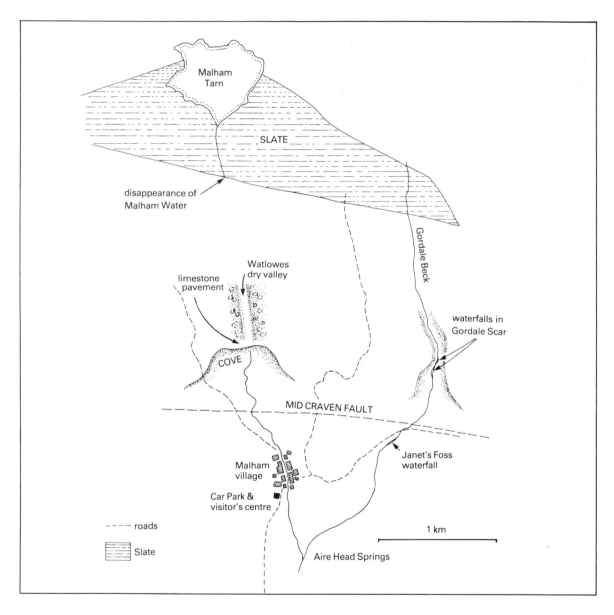

left the brittle strata shattered by networks of fissures and joints. The combination of jointing and solution processes can be seen in any exposed stretch of limestone pavement – like the superb expanse above Malham Cove in the Yorkshire Dales. The solution process allows running water to penetrate and enlarge the labyrinthine fissure network, so that surface streams almost invariably find their way underground quite swiftly. The water may suddenly vanish through an unassuming little fissure, but continuing solution may produce a yawning chasm. Similar swallow-holes may be formed not by the initial burrowing of a river, but by the collapse of underground caverns eroded by a river in the course of its travels along the fissure network. Unfenced swallow-holes, steep scars and deep gorges all make it inadvisable to ramble across unfamiliar limestone country at night, where one can suffer far worse than the twisted ankles which result from clumsy attempts to cross a limestone pavement.

One of the most spectacular swallow-holes is Gaping Ghyll on the flanks of Ingleborough. The great shaft plunges down for almost 330 feet communicating with an enormous vaulted chamber. From here the waters twist this way and that in the maze of fissures for a good mile, before they emerge into the living world as Clapham Beck, which issues from the mouth of Ingleborough Cave. Here one can witness a common cause of gorge formation in limestone country, produced by the progressive collapse of the rocks in the roof above the mouth of the cave. Though not so deep as Gaping Ghyll, Alum Pot, which lies on the eastern slopes of Ingleborough above Selside hamlet, is said to be the most spectacular swallow-hole, pothole or entrance shaft in the whole of Britain. The mouth of the hole measures 131 feet by 33 feet and the depth is 213 feet. The seasonal stream which plunges into the underworld disintegrates into mist and rain before reaching the rocks below, but becomes a glittering cascade of ice in the depths of winter. The waters then pass through the Long Churn cavern networks. Still in the limestone paradise of the Yorkshire Dales there are The Buttertubs, lying on either side of the road over Buttertubs Pass between Swaledale and Wensleydale. Here the geology is mainly composed of shale and sandstone, but wherever the streams encounter a band of limestone they are likely to embark on an underground mystery tour. So it is at The Buttertubs where five potholes stand in line, the two largest being sixty-five feet deep, with detached columns of limestone that are carved and fluted by the solution of the surrounding rock in the acidic rivulets which flow down from the peat and shale above.

Once a river has disappeared on a subterranean journey its route, whereabouts and destination become matters for

> **Some Spectacular Swallow-holes in the Yorkshire Dales**
>
> *Gaping Ghyll*, SD 751727. The swallow-hole can be reached by walking through Clapdale and Trow Gill. A winch may operate for the two ten-day periods culminating in the spring and late summer bank holidays, giving visitors the opportunity to descend the shaft.
>
> *Alum Pot*, SD 775756. This is on the E slopes of Ingleborough, just above Selside hamlet, and reached by a walled track leaving the through road at the N end of the village. A clump of trees on the fell marks Alum Pot.
>
> *The Buttertubs*, SD 875962. These lie directly beside the Wensleydale-Swaledale Buttertubs Pass road; a small parking space is provided. Caution is needed and dogs should be kept on their leads.

speculation, for some of the laws which govern surface drainage no longer apply. Stream courses may cross and recross at different relative levels; floodwater under pressure may surge uphill; the river may carve a great tunnel-like course, only to abandon it as a new joint system is opened up; and abandoned resurgences may be left high and dry. The underground riverside landscape is likely to include cathedral-like chambers; thundering waterfalls, whose roar is never heard by mortals on the surface; ponds or lakes formed where water fills sumps or basins in the stream course; and other sections where the river rushes and corkscrews through the tubes which it has drilled. Where a chamber has been abandoned, then upward-pointing stalagmites as well as the icicle-like stalactites may form. Water which flows through the limestone, dissolving the rock as it goes, becomes charged with lime. But as such droplets drip from the roof of a chamber, so carbon dioxide is released into the air, and the droplet is consequently obliged to precipitate its calcite load, either as a contribution to the cave floor layer or,

Fantastic calcite formations resembling a petrified cascade in Stump Cross Caverns, near Pateley Bridge

where a succession of drips originate and land in the same place for hundreds of years, in the gradual construction of calcite spikes.

Caving is a hazardous activity which novices should not attempt without a competent guide. Special dangers arise when a cloudburst causes a rapid flooding of the cavern network, flooding sumps which cut the escape route. There are, however, a number of show caves where the worst that one may suffer will be a bout of claustrophobia, and where the ghostly calcite formations offer ample compensation. Stump Cross Caverns are at the roadside above Pateley Bridge in Nidderdale on a part of the Pennines plateau which is scarred by the pits and mounds of old lead workings. They offer a cavern walk with some excellent stalactites and

stalagmites and, above, a display centre where a fascinating video explains the underground world and the harsh lives of the 'old men', the miners who first explored the labyrinths. It was also lead miners who, perhaps in Saxon or Norman times, penetrated the cavern networks around Castleton in Derbyshire. Later, some of the miners turned their attention to the 'Blue John', a colourful variety of fluorspar which was a semi-precious stone. Now a sequence of underground caverns and canyons, perhaps carved during one of the prolonged spells of equable climate which separated the different Ice Ages, are accessible by steps and presented to the public as show caves. One of the most celebrated and commercialized of show caves is Wookey Hole in the Mendips, entered through a tunnel left dry when the waters exploited a lower course. Also in the Mendips there are the famous caves of the Cheddar Gorge, where Gough's Cave has been abandoned by its river and now features as a popular show cave.

Streams and rivulets entering the limestone underworld through fissures or potholes may re-emerge unobtrusively in a hillside or stream-bed, or they may make a more dramatic entry at the mouth of a great cave. Castleton can provide an unforgettable introduction to limestone scenery. Just above the planned Norman townlet is the wonderful dry valley of Cave Dale, with the Norman keep of Peveril Castle standing guard on the scar overlooking the gorge. Cave Dale is a fine example of a melt-water channel. It was cut in glacial times when the limestone fissures were still plugged with ice, preventing the normal underground drainage, so that gushing melt-water was able to gouge a surface valley. At the edge of the townlet is the gaping mouth of Peak Cavern, the most majestic cave entrance in Britain, being ten yards in height and thirty-three yards wide. The path up from Castleton passes several

Some Limestone Caves which can be Visited

Stump Cross Caverns, Yorkshire, SE 088635. The caverns were accidentally discovered by lead miners, who were very active in this area, in 1858. There is now a first class visitor centre and shop above the caves, and a fascinating video on the history of lead mining may be shown. The site is above Pateley Bridge on the road to Grassington, with parking available.

Victoria Cave, Yorkshire, SD 836631. The cave has produced important prehistoric and Roman remains; it is not a show cave, but can be visited by the public via a path through the impressive limestone scars above Settle.

Blue John Caverns, Derbyshire, SK 132832. These caves, in the Castleton region of the Peak District National Park, have a show cave which is open to the public. About 1 mile W of Castleton at SK 139827 is *Speedwell Cavern*, with a show cave created by mining but which leads to natural caverns.

Peak Cavern, Derbyshire, SK 148826. A spectacular show cave on the southern outskirts of the townlet of Castleton.

Wookey Hole, Somerset, ST 532480. The popular show cave lies close to the gorge of the youthful R Axe, where there is a large paper mill. Wookey is the emergence of several streams which entered the limestone higher up in the Mendips, and the cave gives access to the underground river. There is a substantial tourist centre.

Gough's Cave and Cheddar Gorge, Somerset, ST 466539. This is the longest and deepest limestone gorge in Britain. Fine views can be enjoyed by walking up from Cheddar village along the B3135 to Horseshoe Bend. Gough's Cave, ST 465539, was opened to the public in 1898 by its discoverer, Richard Gough; the entrance is by the roadside close to the village.

Creswell Crags, SK 534741. The caves lie on the border of Derbyshire and Nottinghamshire on either side of the limestone gorge carrying the B6042. At the NE end of the gorge a side road runs E to an information centre and car park.

The Manifold Valley and Thor's Cave, SK 100543. This area, on the borders of Derbyshire and Staffordshire, contains fine limestone and riverside scenery. Thor's Cave can be reached via a track and path from Wetton village, about ¾ mile to the NW; from the footpath which follows the Manifold valley, or from the path running W from Wetton into the valley – with a steep final ascent via artificial steps. From the cave mouth there are excellent views of the Manifold valley. Several other caves in this stretch of the valley have yielded prehistoric remains.

Kent's Cavern, Devon, SX 934641. The show cave is in the E suburbs of Torquay, above the Ilsham valley. It was occupied at various stages of the Old Stone Age and has proved to be extremely important as an archaeological site, as well as being a popular tourist attraction. Also in S Devon are *Joint Mitnor Caves*, SX 774667, close to Buckfastleigh and Buckfast Abbey, and *Kitley Caves*, SX 576515, just E of Plymouth, where a small complex of caves was revealed when quarrying in the valley of the Yealm and has been opened to the public.

small stream resurgences, and then one enters the great chamber, which accommodated a rope works from the end of the Middle Ages until 1974. Deep inside, beyond the massive chamber of the Great Cave, the so-called 'River Styx' flows along the cave floor before diverting into a side tunnel and reappearing as a resurgence in the gorge outside the cave mouth.

Cave entrances have been of interest to man since long before the days of the tourist or the lead miner. Hunting communities, exploring Britain during the warmer interglacials and in the shorter mild interludes during glaciations, pioneered several occupations of the more southerly lands before the current recolonization began in the last stages of the latest Ice Age.

Caves representing subterranean river courses exposed in the side of the gorge at Creswell Crags, where several of the caves were occupied by prehistoric hunters

Normally the hunting bands must have settled in open camps and modest rock shelters as they followed the migratory herds of mammoth, deer and horses. But when their wanderings took them into limestone country they were able to exploit the inviting caves – even if it was a priority to dislodge a disgruntled cave bear from its comfortable abode. Occupation normally took place around a hearth in the cave mouth, although the interiors of caves were probably used for rituals, and burials were sometimes made in the cave floor deposits.

Among the limestone caves which were occupied at various deeply prehistoric times is Kent's Cavern, a show cave in the suburbs of Torquay, formed in a faulted block of limestone which introduces a morsel of incongruous scenery into the slate and sandstone country of South Devon. The archaeologically important caves and rock shelters of Creswell Crags are set in Magnesian Limestone and are situated at the Derbyshire/Nottinghamshire border. The caves here were originally ground-water

tunnels, but during a period when the ground was frozen a stream succeeded in slicing the gorge which now carries the B6042 through the limestone ridge, thus exposing the caves in either side of the gorge. In much more recent times the bottom of the trough was dammed to create an attractive lake. The caves at Creswell Crags experienced a series of different occupations during the Palaeolithic period or Old Stone Age, and a display centre provides information about the ancient occupants. Other notable prehistoric remains have been found in the Victoria Cave, near Settle in Yorkshire, and the magnificent cavern complex of Thor's Cave. This overlooks the Manifold valley on the Derbyshire/Staffordshire border, a beautiful wooden cleft which includes several valley-side caves which had prehistoric occupations. Neither Victoria Cave nor Thor's Cave is a show cave, though both can be reached after steep yet attractive walks.

Gorges, some with rivers and some now without, are another striking feature of limestone country; their origins can be varied and controversial. The largest is Cheddar Gorge in the Mendips – and the price of its deserved celebrity is measured in the crowds and traffic congestion which it attracts. One notion which used to circulate in geological circles argued that the gorge was formed by the collapse of a gigantic cavern, but now it is generally agreed that Cheddar Gorge was, like so many of its lesser kin, cut by rivers on the surface during glacial periods when the cavern networks below were blocked by ice. If Cheddar Gorge has a serious rival in Britain, this must be Gordale Scar, described later in the chapter.

Not all the rivers of limestone country have subterranean courses, and some of our prettiest and freshest country is found where there is a juxtaposition of surface drainage and limestone scenery. Dovedale in

The enchanting How Stean Gorge in upper Nidderdale, where the beck has sliced through the limestone strata

---

Some Beautiful Limestone Valleys

Dovedale (NT), Derbyshire, SK 147509. The valley can be explored via footpaths which leave a large car park about 1 mile E of Ilam, beside the road running E from the village.

The Skirfare Valley, Yorkshire. Littondale is a lovely side-branch of Wharfedale, noted for the splendour of its spring and summer wild flowers. Leave the B6160 just N of Kilnsey Crag; Arncliffe is an attractive place to stop, and a footpath runs close to and beside the river, upstream to Litton. A footpath also follows the *Pen-y-Ghent* stream, which joins the Skirfare between Litton and Halton Gill.

How Stean Gorge, Yorkshire, SE 093735. The gorge is signposted from the Pateley Bridge-Middlesmoor road in upper Nidderdale, just S of Middlesmoor. The gorge scenery is quite spectacular, particularly in winter when the limestone overhangs may be festooned with icicles. Sound footwear is advisable for those who choose to take the narrow path in the side of the cliff.

Hell Gill Beck, Yorkshire/Cumbria, SD 778963. The gorge is signposted from the B6259; continue directly along the track which crosses the railway and follow it over the ford and onwards to Hell Gill Beck; it continues to the gorge and falls.

Derbyshire is such a place. Here the Dove has been superimposed upon beds of hard reef limestone (in the manner described in the preceding chapter), and it has cut a sinuous gorge, exploiting lines of relative weakness in the rock. This is a much-visited valley, receiving many visitors from the nearby industrial conurbations, and so Dovedale is best enjoyed outside the peak tourist periods. Most of the Dale is owned by the National Trust.

Almost all the rivers in limestone country are liable to disappear completely or else pursue hyphenated courses which are broken by stretches where the river periodically forsakes its bed for an underground diversion. The Skirfare can be quite an imposing river when swelled by spring rains and snow-melt, but in the dry summer of 1984 a long section of the course above Arncliffe in Littondale was bone dry and traceable only by the bleached rocks of the river-bed. Surface drainage becomes more common at the margins of the main limestone beds, where the narrow calcareous beds are sandwiched between other bands of sandstone or shale. These sedimentary rocks tend to support peat beds rather than the dry pasture which flourishes on the limestone, and so they are able to contribute particularly acid slope-wash and ground water. But despite the potency of the water, the limestone beds may be too thin to allow adventurous developments in underground drainage. One of the finest little passages of limestone river scenery can be explored at How Stean Gorge in upper Nidderdale. This is really gritstone country, darker and more wooded than the limestone areas, but the How Stean Beck, a tributary of the Nidd, flows across a band of limestone left here by faulting. Unable to discover and exploit a suitable fissure network, the beck flows on the limestone surface at the foot of a precipitous and finely sculpted gorge.

Another spectacular trench can be explored where the Hell Gill Beck, a tributary of the Eden, forms the Yorkshire/Cumbria boundary. The Beck has such a generous water supply that its efforts have been directed into carving a deep gorge rather than dissolving and widening the fissure systems of its bed. It takes leave of its gorge in style, cascading over a magnificent forty foot waterfall.

Limestone country was once noted for the beauty of its floral displays. Sadly, the modern tendency to spread fertilizers on the lower pastures has rendered these areas uninhabitable by many of the loveliest plants, though small colonies may survive in the untreated upland pastures at the upper limits of the range of many limestone plants. Most of the orchids in the Pennines have either been lost or exist in small, threatened communities. In Cheddar Gorge, trampling and selfish collecting have taken a heavy toll of the rare and exquisite little Cheddar pink, and to see the floral delights of old limestone country in an unspoiled state one must travel to the Burren. May and June are the best months to enjoy the spangled pastures there. Some lovely plants can still be found in the Dales too: the mountain saxifrage, mountain avens and bloody cranesbill. Grass-of-Parnassus blooms beside some streams in Wensleydale, and Yorkshire has its own lovely bird's eye primrose, with the flowers

like pink confetti nodding on slender stems. It can be found flowering beside streams at the foot of Pen-y-Ghent in June and the visitor can follow the public footpath along the Pen-y-Ghent stream.

While the wonders of the Castleton area should not be denied, there is one area of limestone country – the Malham locality in the Yorkshire Dales – which is unsurpassed for interest, beauty and drama. Malham has three great spectacles, the Tarn, the great white cliff of the Cove and the remarkable gorge of Gordale Scar, all within reach of the pretty, if sometimes crowded, stone village, which once served as a collecting centre for the old cattle droving trade.

As a lake set in limestone country where rivers are few and far between, the Tarn is a remarkable peculiarity. In fact only its northern shore is in contact with saturated limestone and, as a result of geological faulting, most of the Tarn is floored by impermeable slates. The Malham Water flows southwards from the Tarn before leaving the slate, encountering limestone, and sinking below the ground. It eventually resurfaces at Aire Head Springs, half a mile to the south of the village, though the Tarn must once have been a natural reservoir for waters which then hurtled onwards to the Cove.

Malham Cove, a curving white wall which drops vertically for 230 feet, is a stupendous, if geologically somewhat controversial, feature. It was, at a time when rivers flowed above the ground, a magnificent waterfall. It was formed by the recession of the falls on a river or torrent of melt-water which plunged over the edge of the Mid Craven Fault. This faultline slices through the landscape just to the north of Malham village. The base of the Cove has probably been trimmed by glaciers, and while no river now cascades over the face of the cliff, a stream gushes from its foot. Above the Cove is the dry valley of the departed river. This is set amidst a glittering

The Nene near Dog in a Doublet Bridge lock

Semi-natural Fenland at the National Trust's Wicken Fen Nature Reserve

The Customs House of 1683 at King's Lynn

A bog oak, dug up and sawn at Wicken Fen in Cambridgeshire

A typical scene in the Somerset Levels, near Fordgate, showing a 'rhine' or drainage channel and a pumping station

An unspoilt setting at Ranworth Broad

Malham, Yorkshire. The main scenic attractions are Malham Cove, SD 900270, Malham Tarn, 1 mile to the N of the Cove via Watlowes dry valley and Gordale Scar, SD 914635. Malham village can be reached by leaving the A65(T) at Hellifield or Gargrave and continuing via Airton and Kirkby Malham. There is a large car park and visitor centre on the outskirts of the village, and it is advisable to use this car park rather than continuing closer to the attractions on narrow walled roads which have few parking places.

Malham Cove, once the face of a mighty waterfall

expanse of limestone pavement, slashed with deep fissures, and perhaps formed by solution along the joints in the rock at a time when it was covered by a skim of glacial drift.

Gordale Scar is another striking and rather controversial feature. Some believe that the plummeting gorge was formed by the collapse of the roof of a vast cavern or pothole, others that it is a river gorge, carved by surface erosion by a river which also thundered down the face of the Mid Craven Fault. The latter theory probably commands most support today. Hemmed in by overhanging cliffs some 165 feet in height, a waterfall is found a short distance inside the mouth of the gorge. Hardy and well-shod ramblers may attempt to climb up the soft, slippery tufa which the waters have deposited on the low cliff beside the falls. Others may deem it wiser to detour to the head of Gordale and walk down the valley. But whichever route is chosen, a stunningly beautiful sight is encountered above the lower falls. Another waterfall is seen spouting through a hole punched in the limestone wall above. The waters fall for thirty-three feet swirling and disintegrating into veils of mist when the wind blows strongly, then charge across a short stretch of rapids before vanishing over the threatening edge of the lower cascade. This is one of the most perfectly composed vistas which the British landscape can offer.

CHAPTER 3
Rivers of the Wetlands

I sing of Floods muzled and the Ocean tam'd
Luxurious rivers govern'd and reclaim'd

(Samuel Fortrey, 1685)

Then on the morrow of St Martin and within the octave of the same there burst in astonishing floods of the sea, by night, suddenly, and a most mighty wind resounded, with great and unusual sea and river floods together, which, especially in maritime places, deprived ports of ships, tearing away their anchors, drowning a multitude of men, destroying flocks of sheep and herds of cattle, plucked out trees by the roots, overturned dwellings, dispersed beaches.

(Matthew Paris, describing the floods of 1236)

The rivers which drain and attempt to traverse broad areas of flat, low-lying ground face special problems, and their wide valleys are ever likely to be transformed by changes in the subtle balance between the land and the sea. This brief exploration of the rivers of the Fens, Levels and Broads and their settings provides a link between the 'natural' world of rivers, as introduced in the preceding chapters, and the man-made river and riverside features, which are introduced in those which follow. In the Fens of Cambridgeshire, Norfolk and Lincolnshire, in the traditional wetlands of the Somerset Levels, and in the Norfolk Broads we meet the far-reaching transformations of rivers and their adjacent wetlands which have been accomplished by man. The pacification of such areas is normally presented as a story of 'progress', yet when we look more closely, the changes do not always appear to have been so beneficial, and the problems posed by Nature appear to have been more resistant than we imagined to the wonders of technology.

The Challenge of the Fenland Rivers

In the flat expanses of the East Anglian Fens we encounter a landscape which seems to lack all the gentle curves, woods, hedgerows and quiet places which one associates with the 'typical' English countryside. Instead, the emphases are on geometry, uniformity and emptiness. The name itself seems oddly inappropriate, for today one must usually travel far to discover even a tiny pocket of wet Fenland, and any wildness in the scene is the bleakness of exploitation rather than the waywardness of undisciplined Nature. Only the tiniest pockets of semi-natural Fen remain. And yet, the intensely farmed and highly organized countryside still preserves many clues to its watery past. The landscape is not without character and the story of the Fens is a fascinating one, meriting the journey into the perhaps unfamiliar realms of geological history which are needed to fathom the meaning behind the scenery. Some visitors find the level plains and vast skies of the region unsettling, for this is a

while others feel braced by the unimpeded winds which roll the surging clouds from one horizon to the other. Meanwhile the rivers flow slowly, but with great deliberation, in arrow-straight courses which cut across the half-hidden patterns of the old river networks of winding uncertainty.

The moist fastnesses which offered seclusion and retreats to Saxon saints, like Guthlac, or patriots, like Hereward, had already experienced a brief but lively geological history of natural transformations, and had known several different episodes of human settlement. Before the Ice Ages, the chalk ridge of the Lincolnshire Wolds extended southwards into Norfolk, running right across the mouth of what is now the Wash, while the ancestors of the Great Ouse, Nene, Welland and Witham flowed through great gaps worn in the chalk barrier, continuing eastwards to add their waters to those of a great Thames/Rhine river system. Gradually, the river erosion widened the valleys and diminished the ridges, until the chalk hills were reduced to hummocks, enabling the sea to break through and complete the erosion during the various deglaciations.

At the close of the last glaciation, here around 12,000 years ago, the rising sea invaded the Wash, inundating an area far larger than the present gulf. But slowly the great sea-swept basin shrank as the tides rolled in with their burden of glacial materials scoured from the North Sea floor. Meanwhile, the great rivers of the East Midlands – draining an area which is today five times as extensive as the Fens – deposited their loads of glacial sands and silt into the shrinking sea, lengthening their courses as the salt-waters receded.

As the shores of the Wash were creeping northwards towards their present position, England, now blessed with a warm and humid climate, was being recolonized by a dense deciduous woodland. Where we now see the seemingly interminable chequer-board patterns of productive agriculture – and where in Saxon and medieval times there were marshes, meres, birch and alder thickets, reedbeds and rich pastures – there was, in 'Mesolithic' or Middle Stone Age times, a great forested plain, with oak, pine and yew. Man had returned to the British scene, and bands of hunting, gathering and fishing folk plied the tortuous rivers and culled the woodland herds of cattle, deer and horses.

The ancient Fenland rivers will have been slow and indecisive. About the time that farming made its debut in England, around 5000 BC, a slight rise in the sea-level had profound consequences for the vulnerable Fenland scene. Rivers which had long been deprived of the gradients needed for a swift and purposeful flow now overran their banks and inundated their broad flat valleys. The river outlets became choked with silt deposited by the tides which surged far upstream. With their lower courses obstructed by banks of silt, the rivers were ponded-back and spread out across their valleys to form great lakes. Forests were drowned, and soon the fallen trunks were engulfed in the rising beds of peat. This peat accumulated because the waterlogged conditions prevented the bacterial decay of the marshland plants which had replaced the forest. Entombed in the acid peat, the tree trunks were also preserved. From time to time the hard, blackened corpses of these 'bog oaks' are exposed as the peat layers waste away: several examples can usually be found at the National Trust nature reserve at Wicken Fen in Cambridgeshire.

As the peat beds rose, so the river networks were further disrupted. New areas were inundated when the rivers blindly explored new avenues for escape between the freshwater morasses and the barriers of tidal silt.

A further transgression of the Fens took place before the 'Neolithic' period or New

Wicken Fen (NT), Cambridgeshire. Wicken village is 3 miles W of Soham and 9 miles S of Ely. The reserve is on the outskirts of the village, at the end of Lode Lane, where there is a car park. A display of the history of the Fens is provided in the visitor centre, while bog oaks are usually to be seen in the reserve; the Warden will advise.

Stone Age had run its course, so that while much of England had now been pacified as farming country, the Fenland peats were sealed by a layer of sticky blue clays, deposited as the storm tides lapped across a landscape of creeks, salt-marsh and mudflats. Human communities retreated to farm on the rich edges of the Fen, doubtless venturing down the meandering creeks to fish and hunt the great flocks of wildfowl. On somewhat drier land in a loop of the Welland near Maxey recent excavations have revealed a complex of Neolithic ritual and ceremonial monuments, including a great 'causewayed enclosure' (a major meeting place), several circular earthen henge temples, and a 'cursus' or avenue.

At quite an early stage in the Bronze Age there was a slight drop in sea-level, small, but sufficient in this low-lying and marginal area to produce yet another transformation of the Fenland scene. Again the rivers were able to advance their courses across the retreating tidal mudflats. In the valleys the brackish regimes yielded to freshwater ecosystems, while on drier ground standing just a few feet higher than its surrounding, alder, oak and pine were able to re-establish their hold on the landscape. In the latter stages of the Bronze Age, however, there was a return to more waterlogged conditions. The gradients of the channels of the aimlessly wandering Fenland rivers were so slight that fine materials eroded from the inland catchment areas were deposited upon the river-beds, while at each high tide silt was swept upstream and settled when the waters receded. It was during the late Bronze Age, around 800 BC, that a great artificial island of timber latticework was built approximately a hundred yards from the shores of a then-existing Fenland mere. By this time England supported a heavy rural population, and turmoils must have resulted from the consequences of climatic changes which reduced the agricultural area.

The elegant waterfront on the Nene at Wisbech

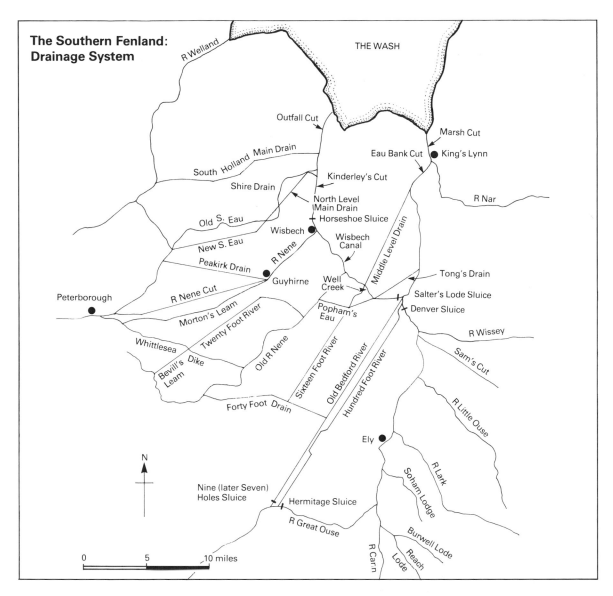

The Southern Fenland: Drainage System

THE WASH

R Welland

Outfall Cut

Marsh Cut

Eau Bank Cut ● King's Lynn

South Holland Main Drain

Kinderley's Cut

Shire Drain

R Nar

Old S. Eau

North Level Main Drain

Horseshoe Sluice

Wisbech ●

Wisbech Canal

New S. Eau

R Nene

Middle Level Drain

Peakirk Drain

Guyhirne

Well Creek

Tong's Drain

Peterborough ●

R Nene Cut

Salter's Lode Sluice

Popham's Eau

Denver Sluice

Morton's Leam

Twenty Foot River

R Wissey

Whittlesea

Old R Nene

Sam's Cut

Bevill's Leam

Dike

Sixteen Foot River

Old Bedford River

Hundred Foot River

R Little Ouse

Forty Foot Drain

Ely ●

Soham Lodge

R Lark

N

Nine (later Seven) Holes Sluice

Hermitage Sluice

Burwell Lode

R Great Ouse

Reach Lode

R Carn

0   5   10 miles

This defensive bastion, over a hundred yards in diameter, was fortified by a wooden palisade. It was discovered in 1983 during the dredging of a drainage dyke near Peterborough, and archaeological excavations began.

As a result of the geographical processes described, the modern Fenland landscape presents evidence of two or three quite different river systems. Much the most obvious is the carefully disciplined post-drainage pattern of straightened channels, artificial drains and sluices. From the air one can still recognize the chaotic, rambling patterns of the earlier days, when the waterways pursued their own haphazard courses to the Wash, collecting hosts of even more wayward tributaries *en route*. Thirdly, there are the sinuous networks of the 'roddons', which survive as slight, winding embankments of paler soil. The roddons, dating mainly from late Bronze Age to Roman times, are fossilized river courses. They consist of beds of whitish silt which

were deposited in the channels of the ancient Fenland rivers. Since the drainage of the Fens, the wasting of the peat soils has lowered the surrounding country, leaving the old river-bed deposits standing above the neighbouring lands. In summer the roddons tend to be masked by the growing crops, but after ploughing, the winding ridges can be quite easily recognized in many places. Though the modern farmsteads of the Fens seem to be sprinkled on the countryside in a thin and random manner, a surprising number of them exploit a roddon situation, and so are less prone to the problems of subsidence which threatens buildings sited on the sagging, sinking peat.

Contrary to popular belief, the Fens did not exist as an untamed and watery wasteland until the great drainage works of the seventeenth and eighteenth centuries.

A typical Fenland drain near Benwick, flanked by fields of black peat

Even so, there were many places which resembled William Camden's description of 1586, of an area near King's Lynn, which was '... so subject to the beating and overflowing of the roaring maine Sea, which very often breaketh, teareth and troubleth it so grievously that hardly it can be holden off with chargeable wals and workes.' In Roman times there was a severe inundation of the northern part of the Fens, but during the Roman occupation the drier lands of the Fen Edge supported a remarkable density of native agricultural and industrial settlements, as well as some villas. The Fenland may have been operated as a vast imperial estate, specializing in livestock production and the evaporation of salt, activities which exploited the extensive grazings and peat fuel reserves of the region. Currently archaeologists are exploring the remains, at Stonea, of what may have been a grand but abortive attempt to establish a Fenland town.

Between the late Saxon period and the great mortality which followed the arrival of the Black Death in 1348 the Fens experienced a remarkable and sustained bustle of colonization and reclamation works. Prosperous towns like Ely, Peterborough, Cambridge, Wisbech and King's Lynn expanded, while in the lands around, the digging of ditches and the building of sea banks (like the still quite well-preserved 'Roman Bank' of about 1250 which rings the Wash) created a landscape in the heart of the Fens in which some of the most prosperous parishes in England were to be found. The task of winning new pastures by excluding the sea from coastal salt-marshes and of draining inland wetlands to create new ploughland and grazing was shared between the various monastic houses and the many peasant communities – with the latter playing the more important role. Though faced with the hazards of malaria and rheumatism, the people of the Fens were able to enjoy levels of prosperity and sufficiency that most other folk might envy. The independence of the Fenfolk, sustained by local resources and able to flourish in isolation, did not impress Camden, who regarded the people as '... rude, uncivil and envious to all others'. But they could afford to stand a little aloof from feudal England, for the Fens provided them with vast expanses of summer grazing – even if one might have to venture by boat to do the milking or resort to stilts for the cattle droving. There were fish and wildfowl in great abundance, peat and sedge for fuel, willows for making traps and basketwork and reeds for thatching.

To the ordinary countryman the Fens must have seemed an attractive homeland, but to the entrepreneur and estate owner its

The Fenland Abbeys. Monastic communities were attracted to the Fens in Saxon times, and following the Norman conquest several prosperous foundations developed. The most striking legacy is the monastic cathedral at Ely, but there are also monastic church remains at Thorney, Crowland, Denney Abbey, near Waterbeach, fragments of Anglesey Abbey (NT), near Cambridge, encased in the later mansion, and the well-preserved gatehouse of the abbey at Ramsey (NT).

potential was compromised by two linked problems. The first concerned the nature of the region as a flat drainage basin for a series of mighty rivers, and the second concerned the effects of the tides, which were ever likely to sweep up silts from the floor of the Wash and dump them in the mouths and channels of the sluggish rivers. The difficulties were compounded at times when the rivers were swollen by heavy rainfall but unable to discharge their waters, for high tides in the Wash caused a reversal of flow and extensive areas were inundated. Writing in 1642, Andrewes Burrell was unable to decide whether it was the sea flooding or the freshwater inundations which constituted the greater threat, but he recognized that the gravest dangers resulted '... when the Sea floods and the Land floods meet, as often they doe, halfe way betwixt the high Lands and the Sea, in that very place like two powerful enimies joyning in one, they doe over-run the Levell and drowne it from one end unto the other.'

Various local efforts to improve drainage and reduce the problems of flooding were made during the medieval period, and the movement gained in intensity during the seventeenth century. On the whole, however, the effect of the works was to transfer the problems of flooding from one locality to the next, for improved drainage in one place increased the flow of floodwater

The New Bedford River at Welney

Fenland landscape at Wicken Fen

entering adjacent, undrained areas. But by the seventeenth century feudalism was yielding to capitalism and the first stirrings of the Agricultural Revolution could be sensed. Funded by 'adventurers' and effected (or otherwise) by 'undertakers', a number of Fenland drainage schemes were introduced in the early part of the century. Many involved Londoners, ready, as ever, to gamble for riches. There was a strong whiff of 'progress' in the air and many Fenland people were deeply disturbed by it. They could point out that the speculators had painted a misleading picture of the region as a hideous morass. But, as always, it was the cottagers' lot eventually to yield to the designs of their masters.

In 1620-21 King James announced that he would sponsor the reclamation of the Fens – in return for 120,000 acres of improved land – and the Dutch engineer, Cornelius Vermuyden, was approached; nothing happened. In 1630 a meeting of the Commissioners of Sewers (a sort of early drainage authority) at King's Lynn favoured

another contract with the Dutchman, but this proposal met with fierce local opposition. Later in the same year, however, the Earl of Bedford agreed to drain the Fens in return for 95,000 acres of land, and he formed an association with thirteen co-adventurers, the Bedford Level Corporation. Vermuyden, who had already worked on the reclamation of Axholme marshes, was employed to undertake the drainage works.

Rather than dredging the channels of the Fenland rivers, Vermuyden favoured a far bolder scheme, one of increasing the river gradients by making straight cuts which would bypass the tortuously winding sections, while sluices would be built to exclude the tides. The work commenced amidst riots by the local people, so that in 1637 local justices of the peace were ordered to supress the disorders. In the following year a gang of forty or fifty men attempted to destroy the ditches made to enclose old common land near Littleport, and rioting affected several other places. Subsequently the greater disorder of the Civil War interrupted work on what was proving to be an unexpectedly difficult and tedious undertaking.

The main component of the original plan was the cutting of the Bedford River, which began near Earith and continued north-north-east across the peatlands for about twenty miles, carrying the floodwaters of the Ouse in an arrow-straight course which bypassed a long detour of the natural river around the Isle of Ely. The Bedford River was embanked, but these defences proved inadequate in times of winter flood. When drainage work resumed a plan was effected to parallel the Bedford River of 1637 by the New Bedford (or 'Hundred Foot') River of 1651, which employed the efforts of over 10,000 labourers. Between the two man-made rivers was a broad strip of washlands which were allowed to flood, diverting the

The Old and New Bedford Rivers. These can be seen from the road crossings of the A142 just to the W of Mepal, between Sutton and Chatteris, or the A1101 just SE of Welney, between Littleport and Upwell. At the Wildfowl Trust reserve near Welney a footbridge crosses the New Bedford R, connecting the visitor centre and car park to the large hide where waterfowl on the Ouse Washes can be observed.

risk of flooding from the country beyond. Now these washes constitute a vital haven for wintering wildfowl, attracting vulnerable species like the Bewick's swan. Near the junction of the Ouse and the New Bedford River the Denver Sluice excluded

A diesel pump of the 1930s, preserved at Prickwillow in Cambridgeshire

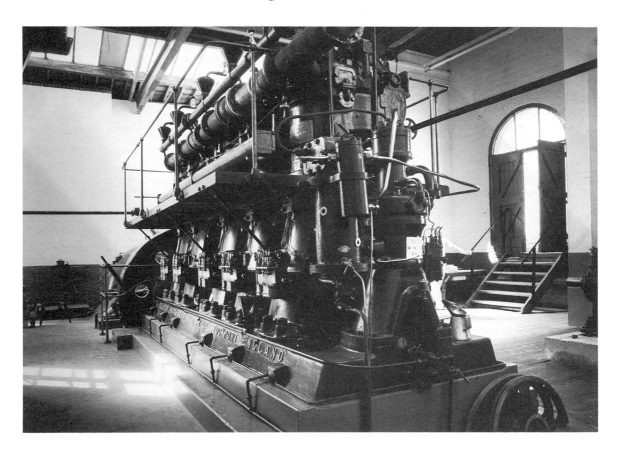

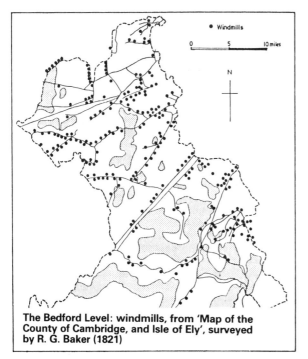

The Bedford Level: windmills, from 'Map of the County of Cambridge, and Isle of Ely', surveyed by R. G. Baker (1821)

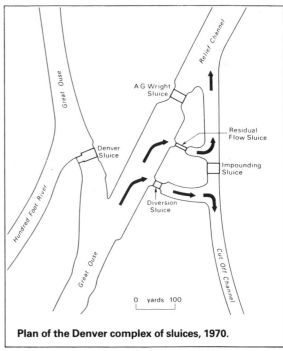

Plan of the Denver complex of sluices, 1970.

tidal water from penetrating the Ouse – necessary to counter the grave mistake of cutting the new river at a level some eight feet higher than that of the Ouse.

One often reads that Vermuyden drained the Fens, but this was true only in an ephemeral sense. Although it was generally acknowledged by contemporaries that he was the best man for the job, the Dutchman's tactical mistakes were overlain by fundamental errors of principle. The main principle concerned was the simple one that when peat is drained and dried, it shrinks: the greater the drain the moister the adjacent ground and the less the shrinkage along its course. As a consequence the main rivers and drains were soon flowing at levels higher than those of the surrounding lands which they were intended to drain. Soon, too, the shrinkage of the peat, which was caused by the removal of water, was compounded by the irreversible problem of wastage caused by the consumption of the black soil by bacteria, which could now flourish in the drier 'aerobic' conditions. To

make the problem still worse, downstream and to the north of the peat fen lay the silt fen, where such wastage was not encountered, so that the peat fen began to form a great sinking basin.

Success was illusory; during the second half of the seventeenth century the newly drained lands produced magnificent yields of grain, beef, butter, cheese, flax and hemp – yet before the century had run its course, flooding had returned. In 1693 the farmers of an area of 30,000 acres around March lamented that 'Where wee should be now plowing the fowles of the ayre are swimming'. Other places, meanwhile, discovered that the drainage scheme was working too well, so that crops were scorched in the summer drought. Vermuyden's scheme had not succeeded, and the quest for progress had proved a hazardous adventure. As Professor H. C. Darby writes (*The Changing Fenland*, Cambridge University Press 1983), 'What seemed a promising enterprise in 1652 had, by 1700, become a tragedy. Disaster

An aerial view of the Parrett, flowing across the Somerset Levels, with geometrical field patterns formed after the drainage of the wetlands *(Cambridge University Collection)*

abounded everywhere.' Now a way had to be found to make Vermuyden's drains work.

Although there were some experiments with horse-powered scoop wheels, salvation was found in the form of the traditional windmill, which could be adapted as a wind pump with a scoop wheel to lift water from one drain to another. But there were also powerful local objections to the introduction of wind pumps. They reactivated the age-old problem of accomplishing the removal of water from one area at the expense of inundating neighbouring fields. Yet there was no practical alternative to the use of mills to counter flooding and, district by district, the disgruntled landowners introduced systems of wind pumps which would lift water from small drains to greater drains, from these to the natural and man-made rivers and on, eventually, to the sea. By the early years of the eighteenth century the wind pumps had displaced the horse pumps, the windmills having received the blessing of authority, in the shape of the Bedford Level Corporation, in 1678. Even so, unco-ordinated mill drainage was to be a curse, and between 1699 and 1716 orders were made for the demolition of more than 120 examples which had proved to be unauthorized nuisances. As this writer has described, 'The mills appeared everywhere, like bizarre dragonflies hatched from the marsh, so that by 1748 there were 250 mills

Windmill silhouetted against a Broadland sunset at Thurne

King's Lynn, Norfolk. Formerly Bishop's Lynn, the town developed on an unhelpful site of old saltings and the estuarine marshes of the Great Ouse, where it grew into a great port, linking the trading worlds of the North Sea and the Fenland rivers, and handled wool, cloth and agricultural exports as well as a range of exotic imports. A legacy of the port function is the Customs House of 1683, built by Henry Bell, with an open ground floor which served as a Merchants' Exchange. Some older merchant houses and warehouses survive, notably the Greenland Fishery House of 1605 and the Hanseatic Warehouse of 1428. *Wisbech* is another distinguished Fenland town, with a fine 18th century waterfront along the N brink of the Nene. The NT preserves *Peckover House* and the early 18th century houses at 14 and 19 North Brink which flank it.

in the middle section of the Fens alone.'

Combined in a reasonably coherent and comprehensive system, the windmills greatly alleviated the flooding problem, but vigorous opposition to the Fenland drainage system was still voiced by the navigation interests. Both King's Lynn and Cambridge depended heavily on river commerce, and claimed that navigation was severely disrupted; the prolonged closure of the sluice doors at Denver was a particular focus of discontent, and the sluice was periodically vandalized by irate boatmen. One problem which defied solution was that of the continuing wastage of the peat, and by the start of the nineteenth century it was evident that the windmills, despite improvements in their design, were no longer equal to the task of raising water from one level to the next. Some new drains, like the Eau Brink Cut of 1821 and the Nene outfall of 1830, were made, but a generally more effective answer was found in the replacement of wind pumps by steam pumps. Although the change had been advocated since the 1780s, the first Fenland steam pumps did not appear until 1817. While such pumps did not use a free and inexhaustible source of energy, they had more muscle, plus the distinct advantages that they would not blow away in gales or be rendered frighteningly impotent at times when the breezes were stilled.

In the middle of the nineteenth century there were probably more than 700 windmills still operating in the Fens between Cambridge and Lincoln, but gradually they were abandoned to decay as the baton passed to steam power. To see a well-preserved example today one must go to the National Trust reserve at Wicken Fen, where the restored and re-erected pump provides a porthole on what was once a common scene throughout the Fens. Steam-powered scoop wheels yielded to centrifugal pumps, then, in the early decades of this century, steam power was superseded by diesel centrifugal

A typical old Broadland scene beside the Waveney, with dairy cattle grazing in the riverside pastures

---

Mementoes of Wetlands Drainage

Wicken Fen Wind Pump (NT). This has been restored to an operating condition and is a distinctive landmark of the reserve: visitors arrive via the visitor centre, cross the footbridge and turn right; the white sails are clearly visible. A later wind pump, built in 1912 on the footings of an older windmill, and working until 1943, can be seen at the 1734 acre *Horsey* reserve (NT) in Norfolk, 2½ miles NE of Potter Heigham and 11 miles N of Yarmouth. *Prickwillow*, Cambridgeshire, with its engine house stands at the former junction of the Lark and Great Ouse, to the NE of Ely at the junction of the B1382 and B1104; the museum is most accessible by water, for craft on the Lark can moor a few yards away.

Denver Sluice. The great modern complex of sluices is at the northern terminus of the New Bedford R, near its junction with the Great Ouse and Wissey. It can be reached by a side road running SW from Denver village, which is on the A10(T) 1 mile S of Downham Market. A fine tower mill stands by the roadside at Sluice Common, between Denver village and Sluice.

---

pumps, and after the last war electrical pumps assumed the burden of Fenland drainage. An imposing specimen of a mighty diesel engine and pump from the 1930s is preserved in the pump house of the 1880s beside the River Lark at Prickwillow in Cambridge, a fascinating museum of the history of the drainage of the Fens.

The threat of flooding was still not completely negated. In the last three years of the 1930s the problem resurfaced, and a severe flood in 1947 served as a warning to the catastrophic inundation of 1953, when a tidal surge was driven deep inland by onshore gales. Attention was again focused on the Ouse and on the ever controversial sluice at Denver, and works were launched to control the discharge of floodwater while stemming tidal invasions. A complex of sluices was then developed at Denver, allowing floodwater to be released into a relief channel, while in dry weather, with the gates shut, the flow is reversed, diverting water away via rivers, tunnels and pipes into Essex.

Further to the west, the Nene has an enormous discharge, exceeded only by that of the Yorkshire Ouse. In March 1947, when snow-melt gushed across the frozen surface of the land to gorge the Fenland rivers, a gauging station near Peterborough monitored a discharge of 285 million gallons per hour. However, earlier dredging and banking operations and the construction, in 1937, of a major lock and sluice near the Dog in a Doublet Bridge passed the severe test of 1947 – though the sluice gates had to be kept fully raised for thirty-one days on end. Meanwhile, on the Welland to the north west the scene resembled a Siberian thaw, with icebergs buffeting and gouging the crests of the floodbanks, while men scurried to plug the gaps with sandbags. One bank failed just to the north of Crowland, and floodwater cascaded through the breach to drown an area of 20,000 acres which extended southwards to Eye and Thorney. Desperate regulation attempts followed, with the pumps at Tydd pumping station being kept working continuously for six weeks. Had they failed at any time, the most terrible catastrophe would have followed. An existing scheme was then implemented, with the cutting of the Coronation Channel around Spalding, the strengthening of embankments and the construction of sluices, weirs and pumping stations.

As a result of these quite recent improvements to the Fenland drainage system the people of the region can sleep more comfortably than ever before. If their dreams are troubled, this is likely as a result of the fact that the wastage of the rich peat still continues remorselessly. On farm after farm the plough is slicing through the last of the moist black soil and bringing up the unwelcome sticky clays beneath. In many places some fifteen feet of peat have been lost since Vermuyden claimed his victory.

The Rivers and Drainage of the Somerset Levels

In Somerset the natural outcrop of Glastonbury Tor stands sovereign over a flat expanse of ancient wetlands. Around this peaty basin, which seems unusual, almost out of place, amongst the mellow, rolling landscapes of the West Country, an arc of hills – the Mendips, Dorset Downs, Blackdown Hills, Brendons and Quantocks – embraces the former lagoon which is now

the Somerset Levels. The origin of the Somerset Levels quite closely resembles that of the Fens, with forests being submerged and beds of peat and silt deposited as climate and sea-level have fluctuated, while quite recent marine transgressions have added their own touches to the scene. Here men have faced the threats of flooding since times immemorial, and the challenges arise both from the rapid discharge into the low basin from rivers like the Parrett, Axe and Brue which follows periods of heavy rainfall in the surrounding hills, and from the very high tides of the Bristol Channel, which pond-back the floodwaters. The drainage problem is underlined by the fact that the gradient on the lower courses of these rivers is often less than one foot in a mile, and it becomes a negative gradient at times of high tide.

In carving their courses to the sea the rivers have cut gaps through a slight coastal ridge, which forms the low northern rim of the basin and presents additional impediments to drainage. The modest hummocks of the Polden Hills run through the moorland basin, dividing the area into Avalon or Brent Marsh to the north east of the hills and Sedgemoor to the south west. In contrast to the low and level horizons of the East Anglian Fens, the scenery of the Somerset Levels is diversified and dramatized by the steep domes of hills like Brent Knoll, Burrow Mump and Glastonbury Tor, cones of tougher rock which have resisted erosion by the ancestors of the Parrett, Axe and Brue.

For thousands of years communities have tapped the rich resources of the Levels and, in their different ways, come to terms with the ever present challenges of flooding and waterlogged ground. Prehistoric communities clearly considered that the wildlife and agricultural assets of the region merited special efforts. The oldest artificial road yet to be discovered is the 'Sweet Track', built well before 4000 BC to link the firm ground

Some Interesting Features of the Somerset Levels

Burrow Mump (NT). The conical hill, crowned by a ruined church, overlooks the village of Burrow Bridge and the Parrett and provides superb views across the drained landscape around. Burrow Bridge is by the A361, between Glastonbury and Taunton.

Ancient Trackways of the Somerset Levels. A reconstructed section of the 'Abbot's Way' of 2800 BC has been built on land behind the E.J. Godwin peat factory, between Westhay and Burtle, accessible from the side road running NW from the B3151 at Westhay. A small museum of the Somerset Levels Project of excavation work is at the *Willows Garden Centre*, 425414.

The Abbot's Fish House, Meare. The swampy lake at Meare was drained in the 18th century and there are few visible traces of the Iron Age 'lake village' excavated here. One fascinating survival from the Middle Ages is the early 14th century Abbot's Fish House, the base of the fishermen of the Abbot of Glastonbury, where fish caught in the mere were salted and stored. Meare lies on the B3151 between Glastonbury and Westhay, and the key for the Fish House can be obtained at the nearby farm.

of the Polden Hills to the island of Westhay. During recent years the archaeologists of the Somerset Levels Project have revealed several sections of artificial trackway with various different prehistoric dates, skilfully built of posts, wattle and brushwood from coppiced underwood, to provide accessways between islands and marsh. A reconstructed replica of the 'Abbot's Way' trackway of about 2800 BC has recently been opened to the public.

Evidence of a later occupation of the Levels came to light after 1892, when the archaeologists Arthur Bulleid and H. St George Gray began excavating Glastonbury 'Lake Village'. The village was thought to have first been occupied around 150 BC and to have consisted of well-built rectangular timber houses which were raised above the marshy ground on oak piles. It was thought that after 60 BC the village was rebuilt as a palisaded settlement of circular wattle dwellings perched on clay floors built above artificial islets of clay and brushwood. Rising water-levels rather than the Roman invasion were thought to have caused the desertion of the village around AD 50. In 1910 the excavators moved to another site near Meare, where adjacent lakeside 'villages', thought to have been occupied from the third century BC to the first century AD, were explored. Little is visible at these Iron Age sites today, though evidence and information are displayed in the museums at Glastonbury and Taunton.

In fact, the question of the 'villages' is less clear cut than this. In the 1960s a new investigation of the evidence brought the suggestion that the presumed 'lake villages' were really prehistoric rubbish tips! A subsequent excavation at the westernmost of the two Meare sites in 1979 raised new questions, and the suggestion that, around 200 BC, the 'settlement' might have been the bogside setting of a large prehistoric market or fair which convened for a few weeks of each year. Evidence of craft industries and vast amounts of discarded material were found, with traces of around 30,000 broken pots.

Like the East Anglian Fens, the Levels experienced a marine invasion in the Roman period, and, similarly, the rich but secretive marshlands offered bases for resistance at different times. One might speculate – rather wildly – that had King Alfred not been able to escape and develop the English resistance in the security of Athelney in the ninth century, then England might now be a Danish-speaking land. The parallel with the Fens was further emphasized by the attraction of the silent marshes and hidden pastures and fisheries to Dark Age and medieval monastic communities. Arriving in search of seclusion, the monks established houses which accumulated and developed vast and valuable ecclesiastical estates. In the course of the medieval period about two-thirds of the area fell into the hands of churchmen, with the Bishop, Dean and Chapter of Wells and the Abbots of Athelney, Muchelney and Glastonbury presiding over their empires of moorland, pasture, ploughland and fisheries. Similarly too, the riches of the wetlands are still proclaimed by a series of exceptionally fine medieval parish churches.

The thirteenth and fourteenth centuries witnessed a flurry of reclamation works, with embankments being built in an attempt to confine the footloose rivers to their channels, while alliances were forged between churchmen and lay landlords to try to rid potential farmland of the threat of flooding. The watercourses of the Brue valley were embanked, dredged and straightened, and the cutting of the Mark Yeo, linking the Brue and the Axe, improved navigation and drainage, while Southlake Moor and King's Sedgemoor were protected by embankments.

Later, however, the energy and enthusiasm for reclamation works became dissipated. By the fourteenth century the climate was deteriorating and sea-storms of increasing fury were recorded around the British coast; the Black Death sapped the strength of communities everywhere, while at the close of the medieval period the Dissolution of the monasteries removed the institutions which had organized and spearheaded the reclamation movement in Somerset. In 1607 some thirty villages are believed to have been inundated by the collapse of the sea wall near Burnham, and many countryfolk and livestock perished. The threat of flooding loomed over the area for centuries to follow; in 1811 Glastonbury became virtually an island of the Bristol Channel, and as recently as 1902 there was a major breach of the sea wall, though today

the region sits more securely behind its sea defences of concrete.

The rivers provided the second challenge in the two-pronged assault of the elements. As recently as 1929 there was a terrible flood when the River Tone, a tributary of the Parrett, burst its south bank and dispersed floodwater across 10,000 acres of farmland, necessitating village evacuations at Athelney Curload and Stathe. Desmond Hawkins has described how one farmer on Curry Moor penned his stock on an island formed by the arch of the river bridge, and fed them there for five weeks until the waters subsided.

The post-medieval fascination with progress and enterprise which directed the destiny of the Fens also permeated the Somerset Levels. Here, however, indecision and sharp practice had a higher profile, and the solidity of intent amongst the greater landowners was less obvious. After the Dissolution Sedgemoor passed from Glastonbury Abbey to the Crown, and in 1618 James I began to coerce the local lords and tenants into supporting a plan for drainage and enclosure of the vast but flood-prone commons. Fortunately, from the commoners' point of view, he was strapped for funds, died in 1625, and nothing was accomplished. In the reign of Charles I the schemes were revived, in a climate of speculation, conflict, intrigue and outrageous corruption. Alder Moor was drained, but many of the walls and ditches or 'rhines' were destroyed by rioters who, like their compatriots in the Fens, realized that 'progress' had little sympathy for the interests of the common man. They could see that their rights in the seasonal pastures were likely to evaporate when the forces of the law and enclosure were unleashed. Here, however, local resistance was sufficient to oppose the projects in Parliament and cause the dreaded Vermuyden, introduced by the London merchant and speculator, Jeffrey Kirby, to abandon his interests in the area.

For a century 'progress' remained

The Halvergate Marshes seen from the Roman shore fort of Burgh Castle, an environment that local agriculture has sought to destroy

dormant, emerging again in the guise of a more fully fledged and muscular agricultural reformism in the latter part of the eighteenth century. This was an age which placed the ordering and disciplining of the environment on pedestals, and abhorred time-honoured practices and untamed countrysides. The tradition, established in Stuart times, that the Levels should attract the most shady types of speculator was perpetuated with the involvement of Lords Bolingbroke and Stavordale, both encumbered by gambling debts. Yet in 1775 the dishonest bill which would have launched them on the untapped riches of Sedgemoor was surprisingly defeated. Then the initiative passed to the more honourable local landowners, who were better able to marshal sound support. The cause of the humble commoners was lost in 1791, with the passage of an Act for the drainage of King's Sedgemoor.

The main component of the plan was the diversion of the Cary from the Parrett, proceeding for twelve miles via a cut across Sedgemoor to join the Parrett estuary below Bridgwater. In 1801 the Brue Drainage Act controlled drainage at the other side of the Levels, introducing a network of new rhines and enclosures, and in the following year a similar Act was imposed upon the floodplain of the Axe.

In these ways a new countryside came into being, one which perpetuates the Georgian view of a neatly ordered landscape. Though largely tamed, the vistas are attractive, with lush meadows defined by straight, willow-flanked rhines. Its pumping stations chugging in some places to raise water from one level to another, the drained landscape of the Somerset Levels has some resemblances to that of the Fens, but it also retains its own special flavour, partly a legacy of the more protracted and piecemeal nature of the drainage operations, and partly a reflection of the fact (recently distorted by subsidies) that the moister western climates favour pasture over tillage.

The region developed as an area packed with mellow character and fascinating details – with cucumbers growing on hillocks of dung, teazles being cultivated for the scouring of fine cloths and felts, willows or withies being grown for basketwork, while wetland plants and wildlife could flourish in the rhines and damp pastures. Farmers in the traditional manner were not entirely averse to flooding, and were pleased to have their fields exposed to river waters in winter. This 'warping' by the silt-laden 'thick water' coated the fields with a fresh skim of silt washed down from the uplands, and once the silt had settled, the floodwater could be drawn away. The ethos of the region became established in scenes redolent of Cheddar cheese and scrumpy, with black and white cattle grazing in meadows which sparkled with a rich variety of herbs and flowers.

The recent developments in agricultural subsidies and grant-aided farming have destroyed many such idyllic scenes, and propelled the region towards the centre of the national stage for conservation disputes and scandal. A confrontation has developed between the local farming interests, anxious to exploit government and EEC grants and incentives by completely draining damp pastures and wetlands – often with the aim of growing otherwise uneconomic but highly subsidized crops – and bodies charged with protecting the national heritage of countryside and natural interest. The conflict achieved a nationwide notoriety in February, 1983, when a group of West Sedgemoor farmers burned in effigy members of the Nature Conservancy Council whose duties they resented. But they sadly misread the climate of popular opinion. In the months which followed the broadened awareness of the threat to the surviving wetlands and river margins which these events had helped to stimulate became focused on incidents in the Norfolk Broads, where the Halvergate Marshes provided the setting for a greater confrontation.

The Norfolk Broads

The Fens and the Somerset Levels share much in common, although differences in the drainage story and the subsequent exploitation of the countryside served – at least until recently – to endow each landscape with its own strong personality. In the Norfolk Broads, too, there are lush wetlands traversed by snaking, lugubrious rivers and straight artificial drains, yet this is a very special landscape – indeed, one that is quite unique. Amazingly, and despite earlier hints and inklings, it was only in 1960, as a result of interdisciplinary researches by Drs J. M. Lambert, J. N. Jennings and C. T. Smith, that the origins of the Broads were fully explained.

The great Broadland rivers, the Bure, Yare and Waveney and their lesser brethren, like the Thurne, Ant, Wensum and Chet, occupy broad, flat valleys. Like the rivers flowing into the Wash, the character of the riverside scene has been moulded by the slight shifts in the precarious balance between land and sea. Here too, phases of waterlogging, occurring at times when relatively high sea-levels have made the rivers aimless and sluggish, have resulted in the deposition of beds of peat. But even during the transgression of Iron Age times the marine invasions here were relatively modest and, away from the lower reaches of the rivers, they were generally insufficient to render the beds of peat inaccessible beneath layers of silt.

Much is still to be learned about the prehistory of the area, but it is known that the Romans exploited the Broadland peat

Burgh Castle. This great Roman fort of the Saxon Shore became a 7th century missionary base and monastic centre associated with St Fursey and later accommodated a Norman motte. There are fine views of Breydon Water by the junction of the Yare and Waveney and the Halvergate Marshes. Reached by water or from the side road running W from the S suburbs of Great Yarmouth. There is parking at Burgh Castle village, ¼ mile from the fort.

beds and the many other agricultural and industrial attractions of the region. In order to defend this distant section of their empire against the assaults of North Sea barbarians they built the great fort of the Saxon Shore at Burgh Castle, part of a long chain of coastal defences and command posts. More recent rounds in the contest between the land and sea have left the fort stranded some distance from the sea, with the magnificent flint and brick walls and artillery bastions standing by the junction of the Waveney and Yare, overlooking Breydon Water.

During the medieval period it was the combination of rich peat beds and the establishment of monastic houses amongst the wetland grazings that produced the phenomenon of rivers which abruptly broaden into expanses of open water which makes the area so distinctive. More curiously, many of the broads do not occur as sudden widenings of the courses of the main rivers, but, like the Wroxham or the Great and Little Hoveton Broads, they lie beside rivers or else occupy tributary valleys. Whatever the motives and convictions which enticed the monks into the quieter corners of the Saxon and Norman countrysides may have been, in place after place the monastic houses would emerge as the focus of vigorous entrepreneurship. Peat diggings or 'turbaries' had a long history in the area, but the monasteries systematically purchased and acquired the

right to dig and dry peat in parish after parish, putting the local people to work in the turbaries. Several hundreds of thousands of turves were exported to consumers in Norwich and Yarmouth as fuel, while the coastal saltings and those of the lower valleys of the Bure, Yare and Waveney consumed enormous amounts of peat for the evaporation of brine. St Benedict's Priory (St Bene't's) was by far the most important of the monastic houses involved, controlling the rights of turbary in a dozen parishes. The industry seems to have been organized in the twelfth century, greatly expanded in the thirteenth, and it continued into the fourteenth century.

Beside the Broadland rivers – which were initially quite similar to other wetland watercourses – great peat diggings expanded. The extraction of the fuel must have been as easy as slicing bacon. The relative sea-level was about a dozen feet lower than today and neat workings could be developed, with vertical cutting faces, clean outlines and uncut ribbons of land left as balks or causeways to preserve tenurial boundaries, serve as flood barriers or as accessways. Towards the end of the thirteenth century, however, the climate and the land/sea balance were again in a state of flux, and the stage was set for the innumerable ecological disasters of the fourteenth century.

High tides and sea-storms began to threaten the lucrative industry, sending surges of brackish water upstream to inundate the workings, or ponding-back the Broadland rivers so that they flooded the

St Bene't's Priory, Norfolk. The modest monastic ruins are dominated, incongruously, by the stump of a much later tower mill. They lie 2 miles SW of pretty Ludham village and can be reached by footpath or by boat, for the site is close to the junction of the Bure and the Ant.

diggings. For a while attempts were made to preserve the industry, with rakes and drag nets or 'dydles' being used to strip peat from the beds of sodden workings. At the start of the fourteenth century some turbaries were still extremely productive. Oulton Broad was producing around 300,000 turves each year, and even in the second half of the century 200,000 turves each year were being produced at South Walsham. Even so, the successive inundations eventually extinguished each monastic enterprise – but they created a unique legacy of riverside scenery and wetland environments, all based on the flooded peat diggings which, by the fifteenth century, had come to be known as 'broads'.

Scarcely had the Broads been created than the gradual transformation of the scene began. At first natural forces played the leading role, but more recently it has been man who has threatened to erase all character from the scene. As soon as the Broadland rivers invaded the turbaries they began to unship their loads of silt, and the accumulating siltbeds at the margins of the Broads allowed reedbeds to become established, stabilizing the deposits and paving the way for the natural reclamation of the Broads by alder carr forest. As a result of these processes, some broads have disappeared and others have become shrunken and much shallower than before. Coypu that escaped from local fur farms in 1937, have stemmed the advance of many reedbeds, but they are generally unpopular, not only because they invade the fields, but also because their grazing and burrowing help to destabilize the river-banks.

The region did not experience a great post-medieval drainage campaign comparable to those which recast the characters of the Fens and Levels. As a result, when tourists and holidaymakers began to discover the Broads in the early years of the nineteenth century they revelled in the watery wonderland. There were rowing and sailing matches, gaudy dance ships, water frolics, pleasure boats and also pageantry, provided by the state barges of the mayors of Norwich and Yarmouth, which met in the Yare each July. The delights of Broadland boating developed in an idyllic setting of reedbeds and lush meadows grazed by dairy cattle, while thatched churches, picturesque windmills and waterside inns punctuated the scene.

Sadly, however, the twentieth century has treated this lovely corner of England unkindly. Pollution from agricultural fertilizers, sewage works, motor launches and farmyards has led to the over-enrichment of the Broadland rivers by phosphates and nitrates. The marginal reedbeds die and the river-banks are exposed to erosion from the washes of the motor boats. The next twist of the fatal ecological knife is provided by the whirling propeller blades, which slice up the surviving waterweeds and churn the polluted waters. Meanwhile, blooms of algae flourish in the over-enriched rivers. One of the algae produces a fearfully effective fish poison. Then, when the algal organisms die and sink to increase the river-bed oozes, they provide an attractive environment for a bacterium which releases a toxin which has devastated the wildfowl population. Yet the impact of man has not always been destructive, for scattered among the Broads are a number of exceptionally valuable nature reserves, some thirty-five of which are managed by the Norfolk Naturalists' Trust, and the Nature Conservancy Council manages the National Nature Reserve at Hickling Broad, the abode of the rare osprey and of the swallow-tail butterfly, which is now entirely confined to two footholds in the region.

Recent years have seen the wetlands of the Norfolk Broads emerge as the leading battleground in the conflict between the forces of conservation and agri-business. Some of these battles might have been

Some Nature Reserves of the Norfolk Broads

Barton Broad, between Wroxham, Stalham and Potter Heigham, was originally bypassed by the Ant. Colonization by reed-swamp and alder carr has been reduced by the grazing of coypu. *Hickling Broad*, in the Thurne valley, is one of the last footholds of the swallow-tail butterfly and noted for its migrant bird life, which includes the spoonbill. *Horsey Mere* (NT), has an unusual brackish water ecosystem, due to the seepage of sea-water from beyond the coastal dunes barrier. It still supports the rare bittern. Fine views can be obtained from the gallery of the NT wind pump. *Ranworth Broad* is the largest of the Bure valley broads, with the Broadland Conservation Centre in the Inner Broad, a bird observation gallery and a nature trail along which the visitor can explore the sequence of reed and alder carr colonization of the Broad's margins.

avoided had the area been awarded the status of a National Park. This was strongly opposed by local vested interests, local government and the drainage authority, and instead a less muscular Broads Authority was created in 1978. The inadequacy of this arrangement was exposed when the Authority found itself unable to raise the funds needed to compensate local farmers who were threatening to drain and plough some of the most valuable parts of the countryside unless bought off with generous 'ecological Danegeld'. The post-war era has already witnessed the destruction by agriculture of about a quarter of the old grazing marshes and their replacement by bland prairie fields. The conflict of interest became headline news in 1984 with the Halvergate Marshes controversy. Recent evidence from the Broads Authority to the House of Commons Environment Committee showed that of 22,000 acres of grazing land existing in 1980, 6000 acres had been ploughed up by 1984. In that year the

Friends of the Earth held sit in demonstrations in an attempt to prevent further ploughing of the Halvergate Marshes, but Professor T. O'Riordan, the Chairman of the Broads Strategy Committee, expressed fears that 1000–5000 acres of grazing would be lost to subsidized cereal cultivation in 1985-6.

A Lucky Escape?

The final section of this chapter does not explore an existing wetland region, but rather an area which, until quite recently, existed as a potential wetland site: London.

As described in the following chapter, the oldest attempt to manage the Thames riverside here that has so far been discovered dates from Roman times. Hitherto, the Thames responded to flooding by dissipating its waters and energies across its floodplain, when the deposition of river silt gradually raised the valley bottom and so limited the severity of the floods. All efforts to control the river deprived the floodplain of these silt accumulations, while the drying and draining of riverside lands lowered their surface by up to a yard – and so greatly enlarged the area which could be inundated should the river ever burst its man-made limits.

Periodically it would do just this, as it did on a grand scale in 1099, in 1236 and in 1242. On each occasion there must have been a severe mortality rate in the capital. In 1663 the whole of Whitehall was flooded, and other notable floods took place in 1791, 1881 and 1928. The near disaster of 1928 should have posted a warning of things to come, and in the terrible east coast floods of 1953 the capital came within a whisker of disaster. The stop-gap measure of raising floodbanks again saved London from inundation in 1965, while two record tides occurred in 1978.

At the national environmental level, the problem was a legacy of the Ice Ages. The accumulation of ice depressed the northern

and western uplands of Britain, which were borne down by the sheer weight of the highland ice-caps and valley glaciers. When the ice melted, the landmass gradually responded by rising. The effect is rather like that of a seesaw with its diagonal fulcrum roughly represented by a line from the Severn to the Tyne: as the north recovers, so the south is depressed and the sea invades the lowland river valleys to produce indented 'ria' coastlines. On the local environmental level, London's problem reflected the drop in the level of the Thames floodplain, already described, and more detailed factors concerning the influence of bridges, docks, dredging and other facets of the man-made river environment.

To concerned officials and planners it must have seemed that a secret alliance of hostile natural forces was conspiring in an attack which would someday overwhelm the capital. In the aftermath of 1953 more earnest attention was paid to the threats. It was recognized that were floodwaters to hit the Underground then the scenes from any disaster movie would pale by comparison to the carnage and destruction. A population of one point two million was dwelling in the floodable zone of seventy-five square miles and, as Stuart Gilbert and Ray Horner have described in their book *The Thames Barrier*, direct damage of the order of one thousand million pounds at 1966 values and indirect

damages of a similar amount would have to be expected.

In the event, London escaped – yet there was a gap of almost thirty years between the uncompromising warnings of 1953 (to say nothing of those of 1928) and the arrival of salvation in 1982, in the form of the Thames Barrier. At any stage during this period a catastrophe of almost unimaginable proportions was possible. The explanations for the delay are set out in detail in *The Thames Barrier* by the aforementioned authors – and the evidence hardly inspires one to conclude that our organizational capabilities and capacity to respond to pressing threats have advanced very far since the days of Vermuyden. Fortunately, the engineering technology is much better. For twenty years vested interests, officials and designers haggled over the location and specification of a barrier, only then could plans be drawn, and it was only in 1973 that tenders were invited. The chosen design has four main navigation openings, 200 feet wide and with rising sector gates to stem incoming floods, and two 100 foot navigation openings with comparable gates, while the free flow of tides is facilitated by four more 100 foot openings, which are fitted with simpler falling radial gates. Three of these are by the north bank and one by the south bank. The operation of all ten gates is from the north bank control tower.

The construction of the Thames barrier, showing the stage reached in June 1980 *(Cambridge University Collection)*

CHAPTER 4

Rivers in Ancient Times

Rivers are among the most obvious and important features of the British landscape, and they have exerted a powerful influence on human activity for many thousands of years. The same river could be both a barrier and a corridor for movement and trading; a boundary or a force for unifying the valley community; a source of food or a threat to farming. The story of river control and management must belong largely to the historical era, but our association with rivers goes back much further.

Rivers in Prehistoric Times

During the glaciations water could flow when the summer snow-melt arrived. Some torrents of melt-water rushed through ice-cut gorges, while other rivers flowed in tunnels drilled beneath the ice, even running uphill under the pressure of accumulating snow-melt. Today the courses of these rivers can sometimes be recognized by the sinuous mounds of sand and gravel which are known as 'eskers' and which grew as sediment was deposited on the beds of the ice tunnels. In some places, where ice plugged many of the old river courses and sealed the mountain basins, great lakes of melt-water were ponded-back until a natural cleft or col could be exploited. Then the waters would gush forth, gouging an escape channel through the rock and flooding the plains below. In deeply prehistoric times hunting bands visited this frigid outpost of Europe during the long, mild interglacial periods and also during the shorter interludes of warmer weather which punctuated the Ice Ages.

As the ice sheets and glaciers waned, human communities returned and explored a countryside which was plastered with drift and strewn with a chaotic assemblage of ice and melt-water deposits, the rivers now picking paths through the scenic carnage of mud slides and sandbanks. With the contraction of the ice cover and the dawning of the Mesolithic period the climate improved rapidly, and the landscape became blanketed in a natural wild-wood. A great pioneering age dawned, with the rivers serving as corridors leading the clans of hunting, fishing and gathering people ever more deeply into the interior of the country. A reading of the Mesolithic evidence of flint scatters and simple camp sites suggests that each inland valley may have served as a clan territory. Winter bases seem to have been established in riverside settings where fishing could supplement the meagre diet of these chilly months. In the spring the families would migrate to exploit the woodland and hunting resources of the valley slopes and higher interfluves, returning once more to the lowland base camps when winter began to bite.

The arrival of farming, around 5000 BC, tended to stem the migratory lifestyle and anchor the communities to hamlets, farmsteads and villages in the agricultural areas. The rich alluvial soils of the better drained riverside settings were some of the first sites to be colonized by farmers, though agriculture soon diffused into other, less favourable areas. We do not know where rivers featured in the religion and rituals of the times; it has been pointed out that some of the great megalithic monuments of the Neolithic and early Bronze Age periods

With a name meaning 'strong' or 'holy', the Ure in its
lower section flows through an area rich in prehistoric
religious monuments

Bolton Priory, with a beautiful situation beside the
Wharfe

seem to have an association with water. Several stone circles, like the Rollright Stones in Oxfordshire or the Stenness Stones on Orkney, are linked with traditions telling that the stones migrate to drink or swim on a certain night of the year. More likely than not such legends are merely superstitions coined in the Dark Ages or medieval period, rather than being reliable evidence of the beliefs of the circle builders.

The evidence for a water-centred religion is much stronger where the late Bronze and Iron Age periods are concerned. The great stone circles, earthen henges and imposing tombs are legacies of the Neolithic and early Bronze Age periods. After about 1500 BC such magnificent monuments virtually ceased to be built, burial customs changed and changed again, while the evidence for any kind of burial custom in the Iron Age is confined to a few special and atypical localities. The negative evidence of the general absence of burials and cremations on the land suggests that corpses were left unburied and exposed to scavengers, or else consigned to sacred rivers and lakes. More positive evidence comes from the numerous late Bronze and Iron Age examples of ornate and costly items of metalwork which have been dredged from rivers and former lake-beds and marshes. They support the notion that (probably amongst other things), people worshipped water gods and subscribed to rituals which involved casting votive offerings to the water spirits and deities. Perhaps the story of King Arthur's sword, Excalibur, cast and then caught by a hand which emerged from the water, brandished and drawn beneath the waves, perpetuated a much more ancient tradition. Amongst the many priceless items recovered from watery places is the circular ceremonial shield of the late Bronze Age, beautifully decorated with a concentric circle motif and apparently hurled into a bog at Moel Siabod near Capel Curig in Gwynedd. A superb bronze sword, splendidly

preserved, has been recovered from the Tay near Perth, and many of the swords of this type which are displayed in our museums have been dredged from rivers, while the Thames has yielded some of the finest late Iron Age exhibits. They include the horned Celtic bronze helmet from Waterloo Bridge; an iron spearhead, richly ornamented in bronze, from Datchet, and the superb bronze shield from Battersea, with Classical decorations half concealing a Celtic two-faces-in-one motif.

Perhaps at the time of the terrifying Roman invasion of the bastion of Celtic fanaticism on Anglesey, or maybe a little before, a votive offering of no less than 150 metal objects was thrown into a lake at Llyn Cerrig Bach. The Celtic religions embraced gods and goddesses which were only loosely comparable to the Classical deities; head and skull cults, and nature spirits. Water also featured prominently in the beliefs and rituals. The spirits dwelling in wells, rivers and lakes were venerated and water was associated with fertility and renewal; often the different elements in the complicated religion would combine. Skulls have frequently been discovered beside wells which were worshipped in Roman-British times, as at the well at Carrawburgh on Hadrian's Wall, which was dedicated to the nymph, Coventina, and the well at Heywood in Wiltshire.

Though not immune to superstitions, the Romans regarded our rivers in a more practical way. From time to time the furthest fringes of amateur 'archaeology' will produce a theory which is delicious in its daftness. A recent example is one which holds that the holloways marking old Roman roads are the relics of Roman canals. Not surprisingly, no evidence of locks is found, and how the canals managed to proceed up hill and down dale is a matter for the imagination! But the Romans did construct a number of real canals, as we shall shortly describe.

Following the return of the wild-wood, Mesolithic settlers will have seen scenes rather like this (the Derwent in Cumbria) as they followed river courses into the interior of Britain

The story of waterways in Britain is far older than Roman times, though only morsels of evidence have been discovered so far. During the Mesolithic period the land bridge linking Britain to the continent subsided, and all subsequent settlers must have arrived by boat. The evidence of Neolithic stone axes, very frequently discovered far from their sources of stone, implies that seaworthy boats were used for coastal trading, and doubtless the prehistoric communities had no difficulty in constructing less robust river craft. Some will have been simple dug-outs, resembling the three log boats found in a quarry site at Holme Pierrpoint and dated to the Iron Age. They must have been washed from their moorings when a former course of the Trent flooded.

We have tended to undervalue the achievements and capabilities of our prehistoric forebears, and proof that a well-developed system of river trading existed in Bronze Age times has been spectacularly demonstrated by modern excavations beside the Thames at Egham. Dredging of the river here frequently yielded quantities of late Bronze Age metalwork, with hints of an associated riverside settlement. Rescue archaeology accompanying the building of the M25 produced evidence of substantial riverside settlements and excavations at the New Runnymede Bridge. The double rows of riverside piles which were revealed probably carried the superstructures of wharves. Three sites explored showed traces of settlement, and if each one represented different parts of one 'riverport' then it would have been a substantial place, extending for more than 320 yards. In any event, the excavations demonstrated that by the late Bronze Age, in the centuries around 800 BC, some rivers were thoroughly exploited, with ports and wharves and artificial bank defences. Prehistoric Egham could have been a significant *entrepôt*,

The Grampian Dee, seen near Kincardine O'Neil, has a name meaning 'holy one' or 'goddess'. The Anglo-Welsh Dee was said to undercut its banks on the English or Welsh side as a means of prophesying which side would next be defeated in battle

exporting the produce of the Thames valley and importing diverse British and continental products for distribution to the interior. So far the site is virtually unique, but other Bronze Age riverports surely await discovery, their surviving traces buried beneath thicknesses of river sediment which have accumulated since those distant times.

Although it is probably safe to say that there were no masonry bridges and cross-country canals before Roman times, we can be sure that prehistoric communities exploited rivers for transport, trade, fishing and fowling. Since the capability to reinforce a waterfront and build wharves as well as seaworthy craft existed in Bronze

Age times, there is little reason to doubt the ability to build timber bridges, though how sturdy and elaborate these might have been we do not know.

Rivers in Roman and Dark Age Times

The Romans are, quite properly, revered for their unprecedented and spectacular achievements in the engineering of roads. They were not blind to the potential of water transport, and by the time of their landing in Britain, in AD 43, they had acquired a sophistication in the building of waterfronts, bridges, aqueducts, canals and water-mills. They introduced to Britain the concept of a unified and integrated development programme which was able to supplant the jealous regionalisms and parochialisms of the old tribal divisions. As members of an expanding but centralized empire which had international trading contacts, the Romans developed London (*Londinium*) at a site on the northern terraces and floodplain of the Thames. London had good connections with the interior of the province and was able to draw on the exceptional agricultural productivity of the English lowlands, and it was also readily accessible to continental shipping. The capital enjoyed such a sustained success that the evidence of the Roman settlement disappeared under the urban accretions of later ages, but excavations in the nineteen-seventies and eighties have helped to reconstruct some of the features of the Roman riverside metropolis.

In the mid-seventies sections of the Roman riverside wall were discovered and explored; it was probably continuous from Blackfriars to the Tower of London. This was a late addition to the Roman town, built in the middle or late fourth century and perhaps a response to the shock of the combined barbarian assault of 367 BC on the province. Hitherto, the river front had been open, but the wall, built with some urgency

and incorporating reused sculptured blocks of stone, answered the threat of waterborne assaults by North Sea pirates and it completed the circuit of London defences. A section which was explored between Queen Victoria Street and Castle Baynard Street showed surviving timber piles which supported a raft of pounded chalk, upon which the masonry wall was built.

In 1981, earlier sections of the Roman waterfront were discovered, including box-like timber structures which might be part of the pier base of the Roman London Bridge. The Rennie Bridge of 1832, which was dismantled and reassembled in the USA, was the successor to a stone bridge of 1209. A timber bridge existed here previously, and the original Roman bridge probably stood in a similar situation. Equally interesting was the clear evidence of Roman riverside wharves, their timber and carpentry still intact and preserved in the waterlogged soils. During the development of the riverside engineering programmes between AD50 and 150, the Thames reached about 110 yards north of its present course, and a succession of later wharves marks the stages in the retreat of the river. In Roman times the level of the river defences and wharves was more than thirteen feet lower than those of today, partly a function of the slow subsidence of the land in the southern portion of England and partly a reflection of a global rise in sea-level from the wastage of the polar ice-caps.

The Roman wharves were rebuilt several times, using massive oaken beams which were skilfully jointed together. Just across the quayside road were imposing Roman warehouses, one with its timber floor still preserved. This warehouse, explored in the vicinity of Old Pudding Lane, had an open front and colonnaded façade, with plastered masonry walls. All these structures testify to the rapid and dramatic transformation of the lower Thames which was wrought by the occupying Romans.

While their London Bridge was a timber construction, the Romans were perfectly capable of building bridges in stone which incorporated keystones in their Classical round arches. Several examples survive on the continent, though the British legacy is modest and probably confined to the traces of low bridge abutments. It has been suggested that the stone bridge at Castle Combe in Wiltshire might be a Roman survival, while Harold's Bridge at Waltham Abbey . might derive from Roman inspiration or be a work of the fourteenth century.

Piercebridge, beside the Tees near Barnard Castle, is a really fascinating place, with the living village sitting snugly inside

The bridge over the Tees of 1789 at Piercebridge stands close to the rediscovered footings of the Roman bridge

Two Roman Bridge Sites

Piercebridge, Co Durham. The village stands inside the earthworks of a very large 11 acre Roman camp beside the Roman Dere Street, now represented by the B6275. It is on the N bank of the Tees at the junction of the B road and the A67 Darlington–Barnard Castle road. The Roman bridge was just downstream of the 18th century Tees bridge.

Chesters, Northumberland, NY911701. Chesters or *Cilurnum* was a large fort guarding the W side of the bridge across the North Tyne. The abutment of the Roman bridge survives across the waters from the remarkable bathhouse of the fort, and can be reached via the bridge near the George Inn, proceeding along the footpath which follows a former railway line.

the rectangular earthworks of the Roman camp. Just downstream of the existing road bridge is an abutment of the original bridge which carried Dere Street across the Tees and onwards to the garrisons which guarded the Roman frontier. Perhaps around the end of the Roman period the force of the Tees destroyed the existing stone piers of the bridge, but the tumbled blocks of masonry were uncovered during gravel digging in 1982 in an area just to the south of the present river course, the river having migrated around 110 yards northwards from the Roman south abutment. The abutment discovered stands about four feet high and has a diagonal corner on the upstream side which acted as a cut-water; an adjacent paved area held the first and second piers and protected them from undermining in times of flood. Masses of fallen stones marked the positions of three more piers, but the northern abutment has been completely demolished by the river. The lack of any arched cut stones suggests that the straight piers carried the road on a wooden superstructure. As the river migrated during Roman times the first two piers were

The Lodes of the Fen Edge

Two of the most attractive of the lodes are those which served the settlements of *Reach* and *Burwell*. Burwell is on the B1102 Cambridge-Fordham road and the lode follows the W side of the townlet, where a few hythes remain. The splendid church is a reminder of Burwell's former prosperity. The green of Reach – formerly two villages – has been cut into the NW end of the amazing Dark Age or Roman frontierwork of Devil's Dyke, and the lode continues this alignment into the Fens and must predate the Dyke. Reach is 2 miles W of Burwell and can be reached from it or by leaving the B1102 at Swaffham Prior village. Both lodes continued to the Cam, about 3 miles to the NE.

redundant and were demolished to build a causeway.

Also in the north east of England, there are the traces of the bridge abutment visible across the North Tyne from the Hadrian's Wall fort at Chesters. Here the river has moved westwards and the bridge abutment is stranded. Three piers supported a road about twenty feet wide, and when the river is low the other abutment and bases of two piers become visible. However, the visitor will be much more impressed by the considerable remains of the stone bathhouse nearby, a reminder of the comprehensive Roman command of water engineering techniques.

The Roman repertoire of water engineering capabilities extended to viaducts, although again the British relics cannot compare with the continental legacy. The existence of aqueducts is evidenced by the baths, water supply pipes and drains excavated at a score of Roman towns. Natural streams were tapped by artificial waterways that were capable of delivering millions of gallons of water to the new towns, and traces of these channels have been recognized at Dorchester, Leicester and Wroxeter. At Lincoln, waters from a distant spring were carried and raised seventy feet to the level of the town by an aqueduct of arches which supported the terracotta pipes. By the crossing of a stream at Nettleham the masonry footings of eleven piers of an aqueduct can be seen.

Some fully fledged canals were also built, the best known being the Car Dyke. This is a rather controversial waterway, said by some to be a means of carrying floodwater to the sea; it was probably built early in the occupation for moving supplies from the bountiful farming areas of the Fen Edge, firstly to the fortress at Lincoln and then onwards to York. The canal, which linked natural stretches of navigable waterway and often cut across the grain of natural drainage, was not entirely a glowing

Car Dyke, Cambridgeshire. Various sections of the Roman canal survive. A long stretch runs just beyond the NE outskirts of Peterborough, where it is crossed by several N-S Fenland roads, like the one from Peterborough to Newborough village. The Car Dyke continues N to meet the Welland near Peakirk, where it is incorporated into the fascinating Wildfowl Trust reserve.

advertisement for Roman engineering prowess. In the south it was fed by a now lost watercourse and sluices and weirs must have regulated the level. But the linking of the Cam and Ouse resulted in absolutely terrible flooding in the third century AD, when the control system must have been neglected. The canal was restored by 270 and sections continued in use by navigation during the Saxon and medieval periods, not always with a lock system in operation. It

A section of the Roman canal, the Car Dyke, in the south of Lincolnshire

can be seen in many places, though is seldom easily distinguishable from a deep but quite modern drainage ditch. It is crossed by a road near its southern origin at Waterbeach near Cambridge, and another good section can be seen at the Wildfowl Trust reserve at Peakirk near Peterborough.

The Fen margins also contain some other Roman waterways, which are known locally as 'Lodes'. These canals supported a very active river trading system in medieval times, and they continued to import the products of North Sea trading and export the agricultural goods of Cambridgeshire until the railway age of the nineteenth century. However, it is generally impossible to distinguish between the Roman lodes and those which are only of a medieval vintage, but at least three and possibly six of the lodes are Roman. Reach Lode is certainly of this date, for the stunning late Roman or Dark Age frontier earthwork of Devil's Dyke clearly exploited this canal as a ready made northward extension of the barrier. During the railway age the lodes became mere drains or venues for pleasure boats, but they added colour and diversity to an area otherwise rendered anonymous by modern farming. Consequently a lively local campaign arose when the authorities proposed their abandonment at the start of this decade.

The Roman influence on our watercourses was not always deliberate or benign. Recent archaeological work in eastern Yorkshire has discovered a great silt-up. It is suggested that the highly profitable agriculture of the period eventually resulted in wholesale soil erosion which caused the rivers here to become choked in silt. Perhaps the cultivation of winter wheat exposed the bare plough soils to the ravages of winter, with the unimpeded slope-wash stripping away the soils and dumping them in the river channels.

During the closing decades of Roman rule many towns were crumbling, and when the

legions were partly withdrawn and then finally evicted in AD 410 the whole infrastructure of civilization gradually decayed. Several centuries passed before societies recovered from the plagues, environmental difficulties, economic decay and political disruption, and began – on a more modest scale – to build or rebuild bridges and reactivate waterways. Nothing of the Dark Age legacy of bridges obviously survives, though much of interest may be revealed by excavations. A good example is the 'Viking' bridge excavated beside the upper reaches of the River Hull at Skerne near Great Driffield. It was discovered in 1982 during the excavation of a new lagoon at a fish farm. Waterlogging had preserved the buried oaken piles which had supported the wooden bridge, but at first they were interpreted as part of a waterfront jetty, until the traces of the approaching gravel

Burwell Lode, approaching the townlet where traces of the old hythes still survive

causeway were recognized. As well as revealing information about bridge building methods in the times of the Viking settlement, the excavations produced strong suggestions that the ancient traditions of religion and ritual had been revived. Too many useful objects had been deposited in the river here for sheer carelessness to provide an explanation. Carpenters' tools recovered included four knives, an adze and a spoon drill, while a ninth- or tenth-century iron sword, still in its wooden sheath, and a spearhead were also retrieved. Also found were a score of animal skeletons – horses, cattle, dogs and sheep – but none with signs of butchering for human consumption. Interestingly, the old peat beds bordering the River Hull have also yielded Bronze Age

Modern Name	Celtic original	Meaning
Avon	abonā	river, water
Cerne, Char	carn	stony river
Clyst, Clyde	cloust-	cleansing one
Darent, Dart, Derwent	deruentā	oak fringed stream
Dee	deva	river of the goddess
Devon, Dowlas	dubo-	black river
Don, Doon	Dēunonā	river of the goddess
Earn	arā	flowing
Eden	ituna	gushing
Ellen	alaun-	holy
Esk, Exe, Usk	iscā	water
Forth	voritia	slow river
Frome	fram-	fair river
Glyme	glimo-	shining water
Humber	humbro-	good(?)
Kennet	cunetio-	regal or holy
Leven, Leam, Lymn	leamhain	elm river
Lochy	lochaidh	dark river
Lossie	lossa	herb river
Nene	nēnā	bright river
Nidd	nido-	brilliant river
Ouse	usso-	water
Oykel	uxello	high
Pant	panto-	valley
Taf	tamos	water
Tamar, Team, Thame, Teme, Thames	tamos, teme-	water or dark river
Tarff	tarbh	(charging like a) bull
Tay	tausos	powerful
Tees	tes	surging
Test	trest-	swift, strong
Trent	trisanton	trespasser: the flood-prone river
Tyne	ti-	water
Ure	isura	holy
Wear, Welland	uisu	river
Wey, Wye	wey	flowing
Wyre	uigora	winding

swords, probably other votive offerings from a period two millennia earlier than that of those found beside the Dark Age bridge.

The Dark Ages and the early medieval centuries witnessed the establishment of thousands of hamlets and villages – almost all of them bearing Old English or Danish names. Over most of England the various old Celtic tongues gave way to Saxon or Scandinavian dialects, yet many of the old topographical names – including scores of river names – survived the transition. And so we recall our British forebears each time we mention a river like the Avon or Usk. A selection of Celtic river names, along with their probable meanings, is listed opposite. Glancing at this list, a number of thoughts may occur. Firstly, the ancient sanctity of water is underlined by the numerous 'holy' names associated with rivers like the Dee, Ellen and Ure. Secondly, some of the names which simply mention 'water' or 'river', like the common Avon, Esk and Ouse names, may suggest that such rivers were named by strangers and travellers, and one might

The Roman waterway of Reach Lode served local commerce until modern times

expect that local communities would have brought a little more imagination into the naming of rivers. So perhaps a river would have a universal name and several local names as well. A very high proportion of river names survived the transition to Germanic languages, and so one must wonder whether, much earlier, some pre-Celtic river names survived the adoption of Celtic tongues. There are several river names which experts have found difficult to ascribe and translate in any known language – like Carron, Parrett and Severn – and it could be that these are such survivals.

The names of major rivers have tended to endure. Some rivers did acquire English names – like the Swale (*swalwe*, 'whirling river') or the Wensum (*waendsum*, 'winding river'). A number of lesser streams were renamed, like the Ray and Ree (*atter ee*, 'at the river') or Idle (*idel*, 'slow'). In some cases the old river name was abandoned and the river took its new name from a notable river settlement, as with the Chelmer, from Chelmsford : Cēolmaer's Ford.

CHAPTER 5
Rivers in Service

Most people are aware that rivers are useful as well as beautiful. If asked, they would probably think firstly of the recreational importance of rivers, as venues for fishing, boating and riverside rambles. Then one thinks of the value of rivers in drainage, domestic and industrial water supply or hydroelectric power. But it is only when we probe a little more deeply that we begin to appreciate the immense importance of rivers in former times, when they served all sections of society in a wide range of indispensable ways.

Rivers and Navigation

Popularly, water transport tends to be associated with the canal age of the earlier stages of the Industrial Revolution: a great flurry of ambitious and ingenious programmes in engineering the environment, which were quite soon overtaken and superseded by the achievements of the railway age. But this is a very abbreviated and incomplete section of the whole story of water transport, for rivers were the commercial lifelines of the medieval kingdom, and the story of their exploitation goes back much further, as we have shown. During the Middle Ages rivers provided a means of moving goods and cargoes through realms which lacked the purposeful central and regional controls which were needed to pioneer and maintain a decent system of roads. While the old Roman system of roads soldiered on, providing the essential land network for much more than a millennium after the withdrawal of the empire, the Roman canal system also

At Ely the Cam/Ouse was diverted to flow beside the cathedral city, perhaps in the thirteenth century

continued in use. Some new cuts were made, some channels were cleared, but every effort was also made to utilize rivers and ditches which, to modern eyes, might seem far too hazardous or narrow to be of any interest to the navigator. The Foss Dyke in Lincolnshire was part of the Roman Car Dyke system which was built soon after the conquest. By the Norman period, and probably before, it was back in use by navigation, being restored in 1121, though it was not again the subject of major

In the late thirteenth century, the Clwyd was straightened between the Edwardian fortress town of Rhuddlan and the sea

improvements until the 1780s, as part of a Trent-Boston-Wash water route.

Although the earliest conventional post-Roman water supply reservoir built in Britain and identified by historians was constructed at Whinhill in 1796 to supply nearby Greenock, the first reservoir recorded dates back to 1189 and was built near Winchester to provide water for a 'flash lock' on the newly made Itchen navigation. These flash locks were commonly provided on the more difficult stretches of the river transport system. They were much less sophisticated than the later canal locks – and were unlikely to attract the faint-hearted. They worked in the following way: downstream of a set of inconvenient shallows a weir was constructed, but its central section – say fifteen feet across – was built as a movable barrier. When craft needed to move upstream, then this section was removed and boats were hauled through using sheer muscle power. Often large labour gangs were waiting to assist the boatmen. Then the barrier was replaced, the upstream water-level rose, and the boat continued over the shallows. But the downstream journey was much more eventful, for when the boat had traversed the natural shallows and reached the weir, the gap was opened and the craft shot through on a sudden surge of pent up water, rather like a cork from a bottle of champagne. The much more sedate passage through locks with double doors, which could be opened and closed in turn to equalize the water-levels on either side, was a later sophistication.

Pound locks of this kind appear to have been introduced during the reign of Elizabeth I on the new Exeter Canal of 1566. Seventeenth-century pound locks have survived on the Wey in Surrey, made navigable under an Act of 1651. The most

primitive flash lock of all was the 'cow flash', accomplished by driving a handy herd of cattle into the river and using their combined bulk as a means of raising the water-level!

While the negotiation of flash locks will have tested the nerve of the seasoned medieval boatman, the average river contained many other obstacles. This was a rather schizophrenic world: the ingenuity of craftsmen and engineers of the period is generally underrated, but they operated in an environment where the ability to organize and integrate activities on a scale larger than the parish or estate was poorly developed. One problem resulted from the use of fish traps. Eels and lampreys were particularly in demand. The traps came in the forms of 'fishery hedges' and the more robust 'kidels', both of which affected river levels and constituted severe impediments to navigation. They abounded in the rivers of the Dark Ages and medieval period. The most important fisheries were those of the

Without water transport, many splendid medieval buildings could never have been built. Some of the stone used in King's College Chapel, in Cambridge, was brought by barge, ship and riverboat from Yorkshire

Fens, and Domesday Book of 1086 lists no less than seventy-seven fisheries in the Lincolnshire Fens alone. Each year Ramsey paid Peterborough Abbey 4000 eels in return for free stone from the Abbey's quarries at Barnack, while the town of Crowland paid the phenomenal sum of £300 each year to the local Abbot in return for fishing rights. As gravel workings expose old meanders of the Trent at Colwick near Nottingham, archaeologist Christopher Salisbury has been able to observe old relics. Domesday Book mentions twenty-one 'piscarae' or fish weirs in Nottinghamshire and between 1973-80 the Colwick workings revealed a Saxon and a Norman example, as well as seven Tudor weirs. The Norman fish weir was built of six rows of oak and holly posts which supported wattle screens. It was funnel-shaped, narrowing downstream and was probably designed to catch silver eels during their autumn migration down the river. The Saxon weir, forty-two yards in length, was built of boulders, posts and wattle and ran obliquely across the river, probably funnelling fish into a wicker basket.

Obviously such structures were a considerable nuisance to boatmen, but in the unco-ordinated world of medieval development the activities of millers often caused a greater annoyance. They impounded rivers with dams and weirs to raise the head of water needed to drive their mill-wheels – and meanwhile navigation might be brought to a standstill, not only at the weir, but also downstream, where the diminution of the flow could cause craft to ground. On some rivers the King granted free right of passage, and here the miller was regularly thwarted as boatmen removed the barrier planks and flash locks and released the miller's head of water. Blows as well as words must have been exchanged on many such occasions. Such scenes were enacted long after the Middle Ages, and even on the mighty Thames it was only in 1751 that Navigation Commissioners were instituted, empowered to build tow paths, remove obstructions and improve locks. The flash locks lingered to the end of the century, with the pound locks at Goring and Cleeve, for example, being built in 1787 to replace the 200 year-old flash locks which were associated with water-mills.

While little was achieved in the construction of artificial canals during the Middle Ages, some stretches of natural river were dredged or canalized and a few cases of actual river diversions have been recognized. A notable example is that of the diversion of the river to flow through the notable monastic cathedral city and trading centre of Ely, a minor modification of the Cam/Ouse which was accomplished before the fourteenth century, but one which allowed

The magnificent Bingley Five Rise flight of locks

an important riverport and merchant quarter to develop. At some uncertain but perhaps similar date the Great Ouse was diverted at Littleport to the north of Ely, an artificial channel directing the waters to the Lynn estuary, so that the riverports of Cambridge and Ely were linked to the water transport metropolis of King's Lynn, and thence to the North Sea trading arena. At the end of the fifteenth century Bishop Morton had a twelve-mile cut dug to take the Nene from Starground near Peterborough to Guyhirn, south west of Wisbech. He had a tower built at Guyhirn so that he could watch the men at work in the distance; it was still standing in the nineteenth century, while Morton's Leam survives as a pointer to the later canal building age.

Another medieval exercise in river straightening has been recognized at Rhuddlan in Clwyd. The town of Cledemutha (Clwydmouth) was founded hereabouts by Edward the Elder in 921, and in 1073 a Norman town was built in the

north-west quadrant of the Saxon foundation. Its ownership fluctuated between the English and Welsh contestants until 1277, when Edward I founded a new town and a formidable castle. The planned town was placed to the north west of the castle (the older settlements lay to the south) and this must have helped to save expenditure, for ditchers were then put to work on the costly task of straightening the Clwyd between the town and the sea. One more medieval effort at river straightening, but one which was not motivated by navigational needs, is recognizable in the earthwork remains at Sulby Abbey in Northamptonshire. Here the river was displaced from the valley bottom and diverted into a thirteen foot deep cutting made in the valley side. The moist valley floor was then converted into a great fishpond, and subsidiary 'stew ponds', where the growing fry were raised, were serviced by sluices and channels which conveyed water from the river. Schemes such as this and the ones which we shall describe at Fountains Abbey demonstrate an

The Norman motte and bailey, the Bass of Inverurie, in Grampian is guarded by the Urie, seen here, and the Don

Warwick Castle, on rising ground beside the Avon

impressive capability in water engineering.

The boats used in medieval and later river navigation were adapted to the particular waters upon which they would operate, and small boatyards and the workshops of shipwrights were attached to most riverports. The craft tended to be long and extremely narrow – the narrower the river or ditches negotiated, the more slender the boats – with shallow drafts and strong keels to withstand the stresses of grounding and haulage. On the widest channels sailing boats could operate. On the Broads around Norwich cargoes moved in square-rigged 'keels', while 'wherries' with four oars provided passenger transport. A later development was the Norfolk wherry, with a slipping keel which allowed the boats to navigate in shallow waters, and an unstayed mast set in the bows, which could be lowered when passing under bridges. The last commercial wherry retired in 1950, but the *Albion* wherry is preserved by a Norfolk trust and may still be seen plying the Broads. Our very recent forebears were also familiar with more robust craft which could engage in coastal trading as well as the navigation of the broader rivers; the Thames barges are remembered with great nostalgia and a number are still maintained in good sailing order. One other type of craft had a very different pedigree; though not noted for its stability, the coracle could spin round on its own axis and was the popular choice of Welsh river fishermen with nets to tend. Its lineage probably extends back to the more boat-shaped skin longboats of the prehistoric era and of the Dark Ages. These boats were linked with the (perhaps mythical) Atlantic voyage of St Brendan. Today the traveller may be fortunate enough to see examples of the few remaining coracles at places like Cenarth on the Teifi.

Despite the smallness of the commercial narrowboats and the shortcomings of the river transport network, inland transport was crucial to the medieval kingdoms, and much work could never have been accomplished without the boatmen. To give just one example, stone used in some of the building of Kings College Chapel in Cambridge were quarried at the limestone workings near Tadcaster in Yorkshire. It was loaded on riverboats on the Wharfe and passed along the Yorkshire Ouse system to the Humber. Then it was transported down the North Sea by coasters to the port of Lynn, and moved on the final stages of its journey by the riverboats of the Great Ouse and Cam trading system.

Much is still to be learned about medieval waterfronts, but recent excavations have helped to colour the picture. Between 1981-4 archaeologists explored the waterfront of Reading on the Kennet. This was a sleepy little place until 1121, when Henry I founded Reading Abbey; he was so

Tidal inlets of Milford Haven enhanced the defences of Pembroke Castle

Coracles of an apparently prehistoric design may still
be seen on the river at Cenarth

The Frome was incorporated into the defences of the Saxon *burh* or fortress town of Wareham

The Caledonian Canal at Laggan Locks

Picturesque landscaping by Repton at the National Trust's West Wycombe

delighted with his creation that he asked to be buried there (d. 1135). The Abbey was built at the junction of the Kennet and the foundation's artificial millstream, the Holy Brook. A great dump of clay with timber revetments was built to protect the river margins just upstream of the Abbey, and this formed the waterfront for two centuries. When the Abbey was re-envigorated in the early fourteenth century a fence of wattle and posts was built as a silt trap in the first stages of reorganizing the wharves. This allowed the river to do the basic construction work, and when sufficient silt had accumulated a revetment of oak posts and planks secured the shore and formed a substantial quay. A warehouse was built on the shore by the Abbey wharves, but was neglected after the Dissolution. Later, after the passing of the Kennet Navigation Act of 1715, the river

Conwy Castle, guarding the mouth of the tidal Conwy

was again embanked with posts and planks, though the townspeople at this time were not impressed – they feared that navigation should end at Reading and not be extended to Newbury.

Throughout the Middle Ages a location on a navigable river was just as much valued as a position astride a railway or trunk road is today. The greater riverports had magnificent fairs where goods from near and far were traded. Professor H. C. Darby notes the Sturbridge Fair, held on the banks of the Cam, traded such goods as Italian silk, Baltic timber and Spanish iron, while the fair at the planned market town of St Ives on the Great Ouse was frequented, around 1300, by the merchants of Bruges, Cologne, Rouen and other distant, almost mythical places. Now, however, to talk of river trading is to recall former glories. Once the commerce has died and the wharves and warehouses are gone, few traces of the old economy may survive. A few lingering hythes and some fine old merchant houses

are all that remain to tell that the Fen Edge townlet of Burwell was a riverport of some significance which traded until the railway age. Not far away is Commercial End, with its name, a hidden pool where boats were turned and a converted warehouse surviving to tell the really observant visitor that this was an outport on the lode network until great granddad's day. There is still less remaining at Nun Monkton, a monastic centre, medieval planned village and river transhipment port at the confluence of the Ouse and Nidd. The great triangular green remains, but there is nothing to tell the stranger about the trains of pack-horses which must have arrived here, carrying the products of the Vale and Dales or distributing goods from far-off places.

Hedon in Humberside is the shrunken remnant of a neatly planned new town and riverport of the twelfth century. It was linked to the Humber by a canalized river, with three artificial canals running off to the heart of the riverport. After a brief period of prosperity it declined and shrank in competition with Kingston-upon-Hull and the port of Ravenserodd near Spurn Head, which was drowned by the sea during the Middle Ages. Also a 'goner' is the once important town and riverport of Torksey in Lincolnshire, a Saxon fortified town or *burh* at the junction of the Trent and Foss Dyke. After the twelfth-century restoration of the Dyke, Torksey flourished, serving Lincoln as an outport, but it declined during the thirteenth century when Boston captured

Lincoln's wool trade and the Foss Dyke deteriorated. The town has vanished completely, though the small village which stands at one end of its site and the Elizabethan 'castle' mark its former position.

The first awakenings of the industrial canal age were signified by the Exeter Canal of 1566, which was followed by the much less well-known Stamford Canal, linking Stamford to the old Welland riverport of Market Deeping. This canal, proposed in 1570 but not completed until 1673, gave two centuries of service before being abandoned and then was rediscovered in 1958. The great weakness of the navigable river network resided in the simple fact that rivers seldom went exactly where one would have liked them to go, whereas custom-built canals could exploit the best possible trading connections. A distinction can be made between canalized rivers, where the usefulness of a natural waterway was enhanced by dredging, straightening and the provision of locks and waterfronts – a familiar concept in medieval times – and artificial canals, which were completely new cuts with flights of locks to cope with the unnatural gradients and with essential back-up systems of reservoirs, channels, water tunnels and aqueducts. There was also a distinction between the navigations where the boats were moved by manpower and the canals with towpaths for horse haulage.

Examples of canalized rivers are numerous and include the Severn, Trent and Weaver, and most of the earlier canals were concerned with forging side-branches, bypasses and short cross-links to enhance what was still an essentially river-based system. Thus the Kennet Navigation Act of 1715 extended navigation on the Kennet, while the Kennet-Avon canal of 1794 linked the two river systems, and the great Thames and Severn networks were linked via Stroud, greatly extending the navigational possibilities. A much smaller and earlier case

Morwellham, Devon. This was an important port on the Tamar in the 19th century, exporting locally mined manganese and copper ore for processing and sending arsenic for the farmers of the USA. Now the old quays lie high and dry and the site serves as an open air industrial museum. It lies 5 miles SW of Tavistock and is reached by the A390 and the Calstock minor road.

Above Tintern Abbey beside the Wye; riverside sites attracted monastic foundations

Below Fountains Abbey, with its remarkably ingenious and extensive exploitation of the Skell

Mill End on the Thames near Hambleden, a beautiful converted water-mill

Glendale Mill on Skye, a thatched mill of the eighteenth century with an iron overshot wheel installed in 1902

of river modification was the Wey Navigation to Guildford of 1653, which improved the waterway with cuts across the necks of several meanders.

As the movement gathered momentum and the engineers gained in confidence and in the support of rich and powerful sponsors, longer and more difficult canals were attempted. Aqueducts were built to carry the artificial waterways across the grain and valleys of natural rivers. There was the Dove Aqueduct near Burton upon Trent, a twenty-three arch structure, one and a quarter miles in total length, which carries the Trent and Mersey Canal over the Dove; the six stone aqueducts built along a twelve-mile stretch of the Brecon and Abergavenny Canal; the 200 yard long Wigwell Aqueduct, taking the historic Cromford Canal across the Derwent in Derbyshire, with a beam engine of 1849 being used to pump water up from the river to fill the canal, or the graceful Rolle Aqueduct which carried the long-abandoned

Torrington Canal over the Torridge in northern Devon. Canal tunnels were drilled right through upland masses, with 'leggers' lying on their backs and 'walking' the boats through tunnels like the Armitage on the Trent and Mersey or the Chirk on the Welsh Canal.

These were all difficult ventures, but in a few places the natural terrain seemed to have been purpose-built for the canal engineer. Thus in the Scottish Highlands a series of elongated glacial lakes – Lochs Ness, Lochy and Linnhe – occupied troughs scoured in the shattered old rocks of an ancient fault plain which traversed the region from north east to south west. When the Caledonian Canal was completed in 1847 at a total cost of £1,311,270 (Telford's canal of 1803-22 proving too shallow), it connected the great lochs and allowed fishing vessels to move between the two coasts while avoiding the long and chancy route around the north coast.

There were other places where a special

level of ingenuity and determination was necessary, and probably the most visually stunning exhibit in the British canal network is the renowned Bingley 5 Rise. The ambitious attempt to link Leeds and Liverpool, and thus the east and west coasts of England, was accomplished on the Yorkshire side of the Pennines by a contour canal cut into the side of the natural valley of the Aire. At Bingley, however, difficult gradients were negotiated, and the canal was lowered to the new working level by a remarkable flight of five locks, with three more built a little 'downstream'. But the story of artificial canals, though fascinating,

Relics of water-powered industry abound at Cromford, by the Derwent, in Derbyshire

A watersplash cuts the main street of Kersey in Suffolk close to the site of the medieval fulling mill. Superb timber-framed buildings tell of the textile-based prosperity

is really beyond the scope of a book on rivers.

Rivers and Defence

The defensive potential of a river as a natural barrier to assaults on one or more sides of a stronghold is quite obvious. Not surprisingly, there are dozens of examples which could be quoted. Water could form the principal element in the defensive scheme, be a helpful adjunct to walls and ramparts, or rivers could be diverted to feed artificial moats. In a few cases where a fortress stood by an estuary, the river might also play a vital role in supplying the besieged citadel – as every Protestant in Derry (Londonderry) will be only too pleased to explain. There were other cases still where, while the river itself was rather incidental to the circuit of defences, the citadel exploited the steep face presented by an old river bluff.

In prehistoric times the emphasis was placed on communal strongholds which occupied hilltops, with the ditches, earthworks and palisades of these Bronze Age and Iron Age hillforts being ingeniously constructed to exploit the local terrain and oblige attackers to advance across the most difficult ground. In the lowlands of Scotland and Ireland water-girt defended homesteads were built on artificial islands of earth, rubble and brushwood, but these 'crannogs' were a feature of still lake waters rather than of rivers. The recent discovery of the hillfort-sized island in the Fens near Peterborough, which we have described, suggests that other large water-ringed strongholds await discovery.

In Saxon times the emphasis on communal defenceworks persisted. In the earlier part of the Dark Ages there were sporadic attempts to refortify ancient hillforts, but under Alfred and his heirs a new strategy for the defence of the English realm against Scandinavian invasion was developed. It involved the creation of garrison towns or *burhs*, most of them sited in lowland situations which offered some potential for economic prosperity and growth. It is notable that the two *burhs* which best preserve their neatly planned Saxon street patterns, Wallingford and Wareham, both exploit the defensive advantages of a riverside site. At Wallingford the ditch and earthbanks abut against the Thames, while at Wareham the rivers Piddle and Frome protect the town from north and south. Stamford also deserves a special mention. A Danish camp was established about 877 on the northern terraces of the Welland, while in 918 it fell to Edward the Elder, who built his *burh* on the southern side of the river, directly opposite. When a bridge replaced the old river fords to the west, the Danish and Saxon settlement nuclei were directly linked, and the town developed on either side of the Welland, with the Dark Age components still preserved in the street pattern.

After the Norman conquest the accent shifted from communal strongholds to dynastic and royal citadels. Thousands of motte-and-bailey castles were built, some large and imposing, but many small and modest, while a much smaller number of more costly stone keeps were built. Frequently a convenient river was co-opted into the defensive arrangements. At Cambridge the massive motte stands by the old Roman nucleus on the north bank of the Cam, exploiting the river bluff and guarding the strategic crossing. Choosing just one more from the many examples of river-guarded motte sites we pick the Bass of Inverurie in Grampian, where the motte and bailey of the Norman overlord stands well preserved in what is now a cemetery. One flank is closely guarded by the winding waters of the Urie, while the castle stands in a peninsula flanked by this river and the larger Don. The Norman builders of several stone keeps also recognized the potential of water-guarded sites. The most obvious

Downstream from Fountains Abbey the Skell is incorporated into the landscaped vistas of Studley Royal Park

example is the Tower of London, originating as an earthwork cast up by the Conqueror in 1066, developed with the 'White Tower' of 1180 and which was completed by William Rufus in the 1090s and secured by two rings of walls by Henry II in the thirteenth century. The castle commanded the riverside from the south-east corner of the old Roman city. One of the finest of the surviving square keeps is at Rochester, perched on the steep southern bluff of the Medway. At Pembroke the cylindrical keep stands at the extremity of a slender limestone ridge which carries the planned medieval town, with the tidal inlets of Milford Haven guarding the northern and western flanks.

During the medieval period many Norman castles were elaborated and enlarged and other new castles appeared. Examples of river-enhanced fortifications abound, including Warwick Castle, on rising ground beside the Avon; Richmond Castle, overlooking the Swale; Goodrich Castle, guarding a vital crossing of the Wye with its moat hacked into a tough rock outcrop atop a river bluff; Carew Castle in Dyfed, overlooking the estuary of the Carew; and the northern frontier castle at Brough, developed by the Normans at the site of a Roman camp, with its approaches sculptured and guarded by a headwater of the Eden. One of the finest natural defensive sites of all was developed at Cockermouth in Cumbria, where the castle was built on a knoll which was guarded on one side by the Derwent and on the other by the Cocker.

An early attempt to protect the approaches to a castle with an expanse of floodable ground was at the very innovative stronghold at Framlingham in Suffolk, built on commanding ground above the Ore at the end of the twelfth century. The most elaborate water defences of the medieval period are to be found at Kenilworth in Warwickshire, Leeds in Kent and at Caerphilly in Glamorgan. The last named castle has the most impressive defences of all, dating from the early fourteenth century, when the outer enclosure was surrounded by an artificial lake while a second lake guarded the northern approaches, these lakes being complemented by a series of banks, ditches and hornworks.

The medieval castle reached the heights of (costly) sophistication with the construction

Some Castles with Impressive Water Defences

More primitive water defences were obtained simply by damming a valley, as at *Saltwood* in Kent, or by raising the castle mound in a swamp, as at *Skipsea* in Yorkshire. Subsequently more sophisticated water defences were constructed, mainly to flood siege tunnels and keep siege and artillery engines away from the walls. The sites were flooded by the damming and diversion of streams, and in the most sophisticated designs the dams themselves were fortified to prevent their being breached by attackers. The most elaborate water defences of all are at *Caerphilly*, Mid Glamorgan, by the A469 and N of Cardiff, built by the Earl of Gloucester about 1268. A 1000 foot long dam-cum-barbican was heavily fortified, fronted by its own lake and reached by gateways and a drawbridge, with another drawbridge giving access to the castle proper. Other remarkable water defences can be seen at *Kenilworth Castle*, Warwickshire, by the A429 SW of Coventry; *Leeds Castle*, Kent, off the A20 E of Maidstone, and *Bodiam Castle* (NT), W Sussex, 12 miles N of Hastings and reached by side roads from the A21 and A229.

of a series of citadels by Edward I, to complete and secure his conquest of Wales. The river works at Rhuddlan have already been described, but other Edwardian castles showed a similar commitment to exploiting the military possibilities of rivers. At Conwy, for example, the whole north-eastern flank of the stronghold, which commanded the mouth of the Conwy, was protected by the tidal river. Admirable though the terrain was, it did pose certain difficulties. In 1297 Edward marched into Wales to put down an insurgency, but a sudden flood on the Conwy trapped the King and his retinue in the castle, severing supply lines so that the garrison was obliged to survive on a diet of bread and decaying meat until the waters subsided and the

reconquest of Wales could continue. It was then decided to build a new castle at Beaumaris on Anglesey: a rigidly concentric castle defended by a sea-fed moat. Across the Menai Straits on the mainland was Edward's great castle and walled plantation town at Caernarfon. The site here was superb, and the castle and its town were carefully tailored to the peninsula defined by the Menai Straits and the estuary of the Seiont. With its fortified quay and sheltered waters, Caernarfon could be supplied and reinforced by the fleet in times of crisis.

Following the gigantomania in castle building during Edward's reign the momentum of activity in the defensive field sharply declined. After the Wars of the Roses and the emergence of the powerful Tudor dynasty the emphasis shifted from internal conflicts and pacification to the defence of the emergent nation against foreign attack. This shift was manifest in the building of Tudor coastal castles at points along the southern coast which seemed particularly threatened, and heavily armoured garrisons and artillery bastions appeared at places like Deal, Walmer and St Mawes.

A trio of relatively late defensive constructions do merit special attention. At Bodiam in Sussex, in an area exposed to French raiding and possibly worse, Sir Edward Dalyngrigge developed a new castle in the 1380s. The site was moated by diverting the Dudwell and some springs to form an artificial lake. At Berwick-upon-Tweed, for long a vulnerable pawn in the medieval wars between England and Scotland, the southern flanks of the town were guarded by the river and the town's old riverside wall. The landward defences of Berwick were greatly enhanced after 1558, with the abandonment of northern suburbs and the building of a great curtain wall with huge arrowhead-shaped bastions of a type much favoured on the continent, but quite unusual in British town defences. These

a The spectacular Humber road bridge of 1981

b The medieval pack-horse bridge at Moulton in Suffolk, one of the finest surviving examples

c The lovely Teston Bridge on the Medway, with Himalayan balsam colonizing the river-bank in the foreground

d The clapper bridge at Postbridge on Dartmoor, with its successor in the background

The outlet of Crummock Water. The water is not very rich, so plants are restricted to common reeds and sedges behind which willows have established themselves

A highland brook in Glenshiel in the Western Highlands

An upland bog in the Lakeland plateau near Eel Crag, looking towards Helvellyn. These bogs develop where water draining off surrounding land accumulates: many mountain streams are born in such environments

An autumnal Lakeland scene near Ashness Bridge in Cumbria. Birch and rowan grow in the rocks by the stream contrasting with the ubiquitous bracken, most attractive in its autumn colours

The Water of Nevis, at the foot of Ben Nevis, is fed by large quantities of run-off water from the mountains in the spring thaw. Pools and riffles that develop in the stony river-bed create highly oxygenated waters

were crowned with earthen parapets in the mid-seventeenth century, making Berwick, with its land and riverside defences, one of the most formidable of military targets.

Despite the great shift in military priorities, London remained vulnerable to foreign waterborne attack. In 1665 a Dutch naval raid on the Medway underlined the dangers in a manner which could not be overlooked, and the engineer Sir Bernard de Gomme designed a new system of riverside anti-invasion defences. The best surviving example is Tilbury Fort on the Thames of 1670-80, with a pentagonal moated core of stone-faced earthen ramparts and four great corner bastions which bristled with gun positions. On the riverside below, two great batteries were placed to guard the Thames. As Christopher Taylor has described, 'Tilbury provides our finest example of the seventeenth century military engineer at his most inspired'. The anti-invasion defences of Britain were revamped following a Royal Commission report of 1860, and Coalhouse Fort was developed near Tilbury, with massive stone-faced earthen ramparts and vaulted casements for artillery. The underground casements were soon abandoned with the introduction of the enormous new breach-loading guns; the stone facings of the ramparts were covered in great impact-absorbing earth banks and open batteries were mounted above.

Rivers and Industry

From the Roman conquest until the application of steam power during the Industrial Revolution, moving water provided the main source of industrial energy in Britain. In prehistoric times grain was ground between the stones of 'hand querns', and it is probable that the Romans introduced the first water-powered mills to these islands. Traces of Roman water-mills have been found on Haltwhistle Burn, the North Tyne and the Irthing, all sites which must have been associated with the garrison

Tilbury Fort, Essex. The original anti-invasion blockhouse by the Thames was built by Henry VIII in 1539. It was strengthened in the reign of Elizabeth I and was the scene of her rallying call in the face of Spanish invasion threats in 1588. These defences were superseded by the star-shaped fort of 1670-80 which experienced the various later refinements described. Tilbury Fort guards the approaches to London at Ford Road, ¼ mile E of Riverside Station.

market for food at the Hadrian's Wall complex of frontier defences. The milling technology must have spread quite rapidly, for it seems likely that scores of simple horizontal mills or 'click mills' were built during the Dark Ages. These were small but effective structures, with a little water-wheel which rotated in a stream in a horizontal position, driving the millstones above via a vertical shaft. In 1980 dates of a dozen excavated click mills in Ireland were published, and they lay in the timeband between AD 630 and 930. Mills of this basic type, however, were used in western Ireland and the Scottish Highlands until the nineteenth century, and a well-restored example can be seen at Dounby on Orkney, straddling a tiny burn.

In Saxon times mills of both the horizontal and vertical type proliferated in England, and excavations at Old Windsor suggest that vertical mills had come back into use by the ninth century, if not before. The earliest recorded mention of a Dark Age mill comes in a charter of 792, granting the use of a mill to a district just east of Dover. Domesday Book shows that at the start of the Norman era there were at least 157 water-mills operating in Sussex alone; the

Dounby Click Mill. This well-restored little mill is on the mainland Orkney island, beside the B9057 about 2 miles NE of Dounby village.

total number of Domesday mills is 5624, and this is probably a considerable underestimate. At the deserted medieval village of Wharram Percy in Yorkshire a Saxon mill dam has recently been excavated, while the restored medieval millpond is now a charming feature of this beautiful and fascinating site. In 1971 excavations at Tamworth, the old Mercian focus, explored a site just above the junction of the Anker and the Tame and revealed parts of the fabrics of two Saxon mills, with traces of their millponds, leats and jointed timbers.

While medieval water-mill structures have not survived intact, scores of later mills occupy the sites of their medieval predecessors, and mills at places like Hambleden on the Thames can be shown to have existed at the time of Domesday. Most medieval mills were provided by the lords of manors – who then took a toll and attempted to outlaw alternative means of milling. Peasants caught milling at home with hand querns faced a hefty fine. Millers, meanwhile, gained a popular reputation for dishonesty, sometimes being allowed to take one-sixteenth of the grain milled, but commonly being believed to take more. As the water-mills became ever more numerous, so too did the problems associated with their use (on the medieval Thames they occurred, on average, at intervals of less than two miles). Not only was navigation obstructed, but also the natural shallows became shallower as silt

accumulated behind the mill weirs. These hazards were reduced if the mill was built on a leat which bypassed the main stream, but even so the repeated ponding-back and drawing off of water caused difficulties both for boatmen and other millers, and various medieval enactments attempted to grapple with the disruptions.

Self-sufficiency was an important facet of medieval monastic life, and the rules of one of the most important orders, the Benedictines, who virtually monopolized English monasticism until after the Norman conquest, made it clear that monastic houses should be situated close to water, mill sites and horticultural ground. Not surprisingly, dozens of houses of the Benedictines and of later orders like the Cistercians, Augustinians and Premonstratensians were built by riversides, leaving us with the familiar and much-photographed vistas at places like Fountains, Rievaulx, Bolton, Buildwas, Jedburgh and Tintern. The monks often demonstrated great ingenuity in their management of water resources – as in the example of Sulby, already given. Rivers could be manipulated to feed fishponds and flush lavatories as well as to power monastic mills. At Fountains, the initially barren and thorny site which was granted to discontented Benedictines was at least blessed by the presence of the Skell, and a mill with two wheels was built over a mill leat here. As it proceeded on its highly regulated way, the Skell lapped the walls of the guest-house, flowed under a bridge and beneath the triple arches of the lay brothers' infirmary, onwards under the lay brothers' refectory, emerging into the daylight to flow beside the walls of the refectory and dormitory undercroft and round the abbot's house, yielding energy and sanitation at different stages of its progress. At Fountains the visitor can explore what is probably the finest surviving example of medieval water engineering, with the Skell still flowing through the twelfth-century works. Leslie

Wharram Percy, North Yorkshire. The deserted medieval village site is 8 miles SE of Malton and reached via the B1248 and a minor road from Wharram le Street village. From the roadside car park one proceeds downwards into the deep chalk valley, past the earthworks of the deserted streets and dwellings on the opposite slope, and on past the ruined church in the bottom of the valley.

Syson records that at least thirty monastic sites in Britain are known to have been served by adjacent water-mills.

As feudalism decayed after the close of the Middle Ages, the milling rights of lay landlords and monasteries tended to pass to freeholding or leaseholding millers, who could no longer rely on captive estate custom or on feudal service for the repair of mills, leats and weirs. In the following centuries the size, efficiency and complexity of water-mills improved considerably, while rises in the productivity of the ploughlands increased the amount of grain arriving at the miller's door. By the nineteenth century mills of a highly evolved form were still exploiting free and constantly renewable sources of energy. However, the development of electrically powered roller mills, the arrival of the railway and the profound improvements in the road system all worked to concentrate the corn milling industry in the towns, and the day of the picturesque rural mill was almost over. Only the growing demand for stoneground flour has helped to preserve the few water-mills which were able to stagger onwards through the days of centralization, amalgamation and closure.

A selection of water-mills of different types can be provided, although space does not permit a comprehensive list of examples of a type of building which is still quite common, if frequently converted to other uses. Dounby on Orkney has already been quoted as a survivor of the old click mill design, and another example is the Troswick Mill at Dunrossness in Shetland. A simple type of vertical mill is the delightful thatched example which harnesses a burn which tumbles towards the strand in the north of Skye. It is an eighteenth-century mill, although its wooden wheel was replaced by an iron overshot wheel which was installed in 1902.

Tidal mills were a special estuarine development which captured the waters of

Houghton Water-mill (NT), Cambridgeshire. The mill is in the charming village of Houghton near Huntingdon, which is signposted from the A1123. It stands on an island in the Ouse, and there has been a water-mill here or hereabouts since Saxon times. A mill here was granted to Ramsey Abbey in 964 and medieval tenants of the Abbey were compelled to use the Houghton mill. The surviving building, of brick and weatherboarding, is of the 17th and 19th centuries and much of the old machinery survives.

the rising tide and then released them to drive the wheel as the tide fell. The well-restored tide mill at Woodbridge on the Deben in Suffolk was built in 1793 and its seven and a half acre pond is now a marina, while the Eling tide mill in the New Forest was restored and opened to the public in 1980. At Carew a tide mill has operated on the estuary just below the castle since 1558, and the present mill, restored in 1972, was in commercial use until 1958 and is now open to visitors. The more conventional vision of a water-mill, as a pretty structure with whitened weatherboarding and jutting sack hoists, is exemplified by East Anglian mills like Elsing Mill on the Wensum, Narborough Mill on the Nar, Corpusty Mill on the Bure and Houghton Mill on the Ouse. The latter is a superb example, though with plain weatherboarding, and now serves as a youth hostel. Hambleden Mill on the Thames between Henley and Marlow epitomizes the picturesque vision of the rural mill. Lode Mill, its pond and flood control sluice still in place is a National Trust property beside Anglesey Abbey near Cambridge, and other Trust properties include the water-mill at the bottom of the garden at Bateman's in Sussex; City Mill in Winchester; Shalford Mill in Guildford; Burnham Overy Mill in Norfolk; Cotehele Mill on the Cotehele estate in Cornwall; Bourne Ponds Mill near Colchester;

Constable's famous Flatford Mill near East Bergholt; Houghton Mill; and Nether Alderley Mill near Alderley Edge; while the National Trust for Scotland owns Preston Mill on the Tyne near Dunbar.

Not all the mills which operated in the medieval kingdom were employed in the grinding of corn. Early in the medieval period manufacturing industry had an urban bias, with the English textile industry being by far the most important form of manufacturing. However, the rapid expansion of most towns which followed the Norman conquest led to serious local pollution in those days of minimal sanitation. Particularly with the introduction of fulling mills in the thirteenth and fourteenth centuries, the textile industry drifted into less populous rural water power sites. At the delightful village and medieval textile centre of Kersey in Suffolk, for example, is a fulling mill operated by the water splash which bisects the main street. These mills used a wheel to harness the current and the power was conveyed to a beam fitted with hammers which beat the woven cloth in the water, driving out the grease and grime. For reasons which are not entirely clear, the widely dispersed rural textile industry of the south and Midlands later tended to become reconcentrated in Yorkshire, where millers were particularly responsive to each innovation in the mechanics of water power, and where an abundance of clear and sprightly streams tumbled from the Pennines. At first the mills were established in places which could be quite lofty and remote, but which were

Fountains Abbey (NT), North Yorkshire. Arguably the most impressive medieval monastic site in Britain with traces of all the stone buildings of the Cistercian house from its foundation in 1132 to the Dissolution. The abbey is well signposted and lies 4 miles SW of Ripon, just off the B6265.

Water-mills Preserved by the National Trust

In addition to the Houghton Mill described, the Trust maintains a number of other fine examples: the working water-mill at *Bateman's*, East Sussex, ½ mile S of Burwash; *Winchester City Mill* of 1744 at 1 Water Lane, at the foot of High Street in Winchester; the early 18th century mill on the Tillingbourne at *Shalford*, 1½ miles S of Guildford on the A281; the manorial water-mill and cider press on the *Cotehele* estate in Cornwall, on the W bank of the Tamar, 2 miles W of Calstock; *Flatford Mill* in Suffolk, on the N bank of the Stour, 1 mile S of East Bergholt; the partly Elizabethan *Nether Alderley Mill* in Cheshire, ½ mile S of Alderley Edge; *Burnham Overy Mill* on the Burn, ½ mile NW of Burnham Overy Staithe in Norfolk; *Bourne Ponds Mill*, a remarkably ornate building of 1591 lying 1 mile S of Colchester in Essex, just E of the B1025; *Quarry Bank* cotton mill and Styal village in Cheshire, 11 miles SE of Manchester on the B5166; and *Wellbrook Beetling Mill* in Co Tyrone, by the B159, 2½ miles W of Cookstown, which was used in the 'beetling' or pounding of linen, which gave the cloth its sheen.

furnished with vigorous and reliable rivers and streams. Subsequently, however, the arrival of steam power sucked the industry down into the swiftly growing coalfield towns.

Relics of the old water-powered textile industry are numerous, though often much decayed. At the St Fagans open air museum near Cardiff the Esgair Moel Mill, one of the last water-powered fulling mills to remain in operation, has been rebuilt and put back to work. The versatility of water power is apparent at Talybont near Aberystwyth, where a mill which was used in smelting operations at the local lead mines produces Welsh tweeds for the tourist market. One of the most impressive water-wheels to survive is the great pitch-back wheel of the Foster

early development of the iron industry, for around 1200, water-powered hammers were introduced for beating the hot ore bodies into shape at scores of little rural forges. The Wealden industry, previously tied to sources of iron ore and charcoal, migrated to streamside sites where hammer ponds were excavated to accumulate the heads of water needed to drive the hammers. This tendency was emphasized towards the end of the fifteenth century, with the introduction of blast furnaces which produced a much greater heating of the ore through the adoption of water-powered bellows. Furnace ponds were excavated beside many more streams, most of them already frequently punctuated with hammer ponds. In any walk through the now overgrown coppices of the Weald one is likely to stumble across mounds of furnace slag, wet or dried-up forge and hammer ponds, troughs which mark the old leats and the nicks which once held sluices. Just to the north of Cowfold is Suffold hammer pond, one of the more accessible examples from the multitude of hammer and furnace ponds which dimple and splatter the face of the Weald. At the other end of Britain, one of the best surviving charcoal smelting sites is at Bonawe, beside Loch Etive, where ore imported from Cumbria was smelted and where a wheel pit remains.

Today it is difficult to appreciate the former importance of water power. The few mills which survive in working order seem quaint left-overs from a dim and distant past. Yet were we able to step back in time across just two centuries then we would find that water power was as essential to civilized life and progress as electrical power is vital today. Even so, we tend to imagine that in a place like Ambleside in Cumbria, now a busy little focus of the Lakeland tourist industry, life was once quieter and free of bustle – yet in the nineteenth century the townlet echoed to the clatter of bobbin, corn, fulling, bark, cotton and paper mills.

Water gardens in the oriental manner at Cliveden near Windsor

Beck Mill near Pateley Bridge in Yorkshire. It was used in the local hemp industry which was active here until the 1960s, in a dale which is packed with the fossils of water-powered hemp, cotton, silk, linen, flax, wool, corn and lead smelting mills. Yet another application of water power is represented by the breastshot wheel in the entrancing Cotswolds village of Lower Slaughter, for the mill is attached to a bakery and the water power harnessed for dough-making. In Co Tyrone the National Trust preserves the linen 'beetling' mill at Wellbrook.

Water power also greatly influenced the

all powered by Ambleside's becks. When we look more closely, we find that the tourist facilities have deveoped amongst and over the old mill races. If we should tend to scorn the 'primitive' nature of this energy source then not only do we undervalue the ingenuity and craftsmanship of the old millwrights, but also we overlook the facts that water energy itself produced little if any river pollution, left the atmosphere pure and tapped a free and infinitely renewable source of natural power, while at the same time the mills seldom detracted from the charms of the rural scene. In these days of Three Mile Island, acid rain, river pollution, rising costs and fossil fuel depletion, any obituary of the water-mill might prove to be rather premature.

Rivers and Recreation

During the nineteenth century the British middle classes began to revel in the varied delights of 'messing about on the river', and today some 30,000 pleasure craft ply the Thames alone. The Henley Regatta dates from 1839; races were then rowed three abreast through the luxurious litter of spectators' skiffs – while the umpire was mounted and pursued the competitors along the towpath. But the enjoyment of the sporting and aesthetic delights of rivers has a much longer history than this. In 1653 Izaak Walton produced his *The Compleat Angler*, describing how there was more to angling than distressing the fish and gaining a cheap meal. Walton was born in Stafford, worked in London, retired to Shallowford in his native county and fished in many places, such as Dovedale and Beresford Dale, before moving to Farnham and dying in Winchester. In the course of his travels he will have become well aware that the recreational possibilities of water included much besides boating and fishing. The fashionable attraction of water must be seen in a broader context: that of the obsolesence of the feudal castle and the rise in its stead of the stately home. Gradually status came to be expressed in comforts, opulence and the creation of idyllic settings, rather than in turrets and private armies.

No genuinely medieval garden survives, although earthworks at abandoned castle and manor sites and contemporary descriptions help to answer some of our questions. Most gardens were small, enclosed and sheltered, with raised flowerbeds, neat paths, terraces and small lawns. Water was an important part of the carefully orchestrated scene, with artificial pools, islands and streams which were fed by springs and canals or created by the diversion of natural watercourses. On a more ambitious scale, high level streams could be tapped to produce the pressure needed to power fountains.

Landscaping on a grander scale appeared in the Elizabethan age, when water pageants of the most extravagant kind were staged. At Elveham in Hampshire a crescent-shaped lake, with islands in the forms of a spiral mount, a fort and a ship, was constructed to entertain the visiting Queen with a lavish water spectacular in 1591. Lyveden New Bield in Northamptonshire, also Elizabethan and a National Trust property, may look like a semi-ruined stone mansion, but it is really a remarkable summerhouse. It stands in its moat amidst the earthworks of an abandoned garden in which canals made a major contribution to the fashionable scene. All the skills in water engineering acquired by the builders of castle moats, weirs, fishponds and mill leats were exploited by the pioneer landscape architects, and as gardens became larger and more elaborate, water, brought in from the natural streams and springs to fill the neatly ordered pools and canals, maintained its prominence in the man-made vistas.

Until the middle of the eighteenth century the accent was unmistakably on a formal, geometrical arrangement of garden features. The arbiters of taste had little time for the

Water-meadow beside the Yeo at Sherborne in Dorset

natural landscape – regarded as a wild and uncouth place. Instead, the followers of fashion sought to impose their will upon the scene, with the beds, terraces, low hedges, ponds and canals conforming to precise, symmetrical patterns, all proclaiming the triumph of the rational and cultivated human mind over the barbarity of the countryside. In the park which lay beyond the gardens and its ha-ha boundary, the avenues, tree mounds and spinneys repeated the geometrical theme. It was the rise of Romanticism which elevated the river to a new eminence in the manipulated and manicured landscape. While houses remained, for a while, quite Classical and symmetrical, the new vogue required that parks should flow quite unimpeded right up to the walls of the mansion, while the parkland scene should resound to the beauties of Nature. In this formulation, Nature was associated with flowing lines, sinuous curves, great expanses of pasture, scalloped acres of woodland and isolated spinneys and tree scatters – all carefully combined in a relaxed and harmonious manner.

Of course, the real countryside of Britain would not readily conform to this stylized vision of Nature, and represented no more than 'capabilities'. Naturalization was only achieved at a very considerable financial cost. It involved the redrafting of river courses, the construction of enormous dams to form artificial lakes – which were preferably of an indented and winding shape – the construction of waterfalls and ponds and the shifting around of gigantic masses of earth to create a tastefully contoured terrain. Considerable skills were needed to survey and execute the complex forms and levels, much raw labour to rearrange the landscape, and great wealth by those who footed the bills. William Kent (1685-1748) established basic themes which were developed by Lancelot 'Capability' Brown (1716-83) and Humphrey Repton (1752-1818). Rousham in Oxfordshire was one of Kent's most influential creations, with the lawn-like pastures running down to the winding Cherwell.

West Wycombe Park, with its swan-shaped lake filling the valley floor, is one of Repton's works. A most remarkable naturalization of the landscape was accomplished at Stourhead in Wiltshire in 1741 by the architect, Henry Flitcroft, and his employer, the banker, Henry Hoare. The unexciting valley of an obscure headwater of the Stour was dammed just below Six Wells Bottom and transformed to accommodate a chain of irregular lakes. The flanking slopes were planted with woodland, while Classical bridges, temples and grottoes were scattered tastefully about. In the nineteenth century the park acquired some new touches, emphasizing the current fashion for exotic trees and shrubs and colourful splashes of blossom.

At Blenheim Palace in Oxfordshire another undistinguished river, the Glyme, achieved magnificence. The palace of 1705-22 was a national gift to the 1st Duke of Marlborough, designed by Vanbrugh, with gardens by the royal gardener, Henry Wise, all in the formal French manner of Le Nôtre. With the change in fashionable taste Capability Brown was invited to naturalize the scene. This he did by introducing sweeps of woodland and pasture, while damming the marshes of the little Glyme to create a magnificent lake. In 1925-32 the French garden designer, Duchêne, recreated a series of formal water gardens in the early eighteenth-century manner, its parterres, ponds, fountains and intricately symmetrical box hedges overlooking the lake and providing a striking contrast between the juxtaposed visions of perfection.

If Fountains Abbey boasts our best medieval monastic waterworks, the section of the Skell immediately downstream displays some of the finest accomplishments of the eighteenth-century landscape gardener. Studley Royal has water gardens which echo the older, more formal ideas of an ideal setting. They were created by the owner, the notorious speculator, John Aislabie, in 1716, with neat lawns and the geometrical Moon Pool and its fountain and lakeside Temple of Piety. Moving upstream, these water gardens adjoin the Fountains Abbey Gardens, so that one can pass directly into the Romantic. This part of the valley was acquired by Aislabie's son and landscaped in the newly fashionable 'natural' manner, with the less disciplined river flowing between woods and lawns to a lake, with the Abbey ruins forming a magnificent eyecatcher.

It was in the nature of things that, as the fashions in landscaping changed, the latest vogue would obliterate much of the previous designs. At Chatsworth in Derbyshire, however, four phases of landscape manipulation and titivation can be

a

b

c

Freshwater fish can be loosely grouped into types found mainly in the different reaches of a river: *a* dace favour clear-running waters; *b* pike and its prey; *c* rudd and *d* roach, prefer slow-flowing waters. Solitary inactive lives in sluggish, still waters are led by *e* bream. Bream shoal during the breeding period

d

e

River Thame near Dorchester. The river's eutrophic waters support flourishing clumps of submerged, emergent and bankside vegetation such as stonewort, yellow water–lily, lush clumps of marsh–marigold, hemp–agrimony, willow–herb and willow

discerned in these gardens of the Derwent valley. From George London's formal gardens of 1688, which had canals, fountains and an orangery, a canal and the fountain of Neptune survive. Water preserved its command of the scene when Thomas Archer's exceptional great cascade was added a little later. Capability Brown then undid much of the work of his predecessors, naturalizing the valley scene and creating a woodland park beside the river. In the early nineteenth century Joseph Paxton transformed the scene again, introducing fashionably exotic conifers and rockeries as well as conservatories which had since been lost. A striking addition was his Emperor Fountain, with its 296 foot high jet. Although parks and gardens often experienced total transformations, parkland could sometimes preserve archaeological features which would otherwise surely have been destroyed. This is demonstrated at Anglesey Abbey, where a fine set of medieval fish ponds endure as earthworks in the lawns.

Water-meadows

As well as being valued by the navigator, military engineer, fisherman, miller and landscape gardener, rivers were also of use to the farmer. Apart from their obvious uses as watering places for livestock, sources of fish and wildfowl and, quite often, of drinking water too, rivers played a crucial role in the pastoral economy. Attempts to regulate the drainage of flood meadows has a long history, and at Spofforth near Wetherby one can recognize the medieval banks built by the local priory to enclose meadows beside the Crimple Beck. Fully developed water-meadows provided an intricate system of lowland management in the seventeenth and eighteenth centuries, and though the old techniques have almost entirely been abandoned, they have left a rich ecological and scenic legacy.

Lowland sheep farming enjoyed a boom in the seventeenth century and early eighteenth century, but the capacity of the industry was constrained by shortages of fodder in the coldest months. The problem was met by flooding valley bottoms early in the year, thus stimulating a flush of growth. Sometimes the fields, which had been manured and enriched by sheep during winter and spring, would then be cultivated for grain, in other cases the sheep were removed in April and the meadows were again flooded to produce a hay crop in June. The technique was most popular in the chalkland areas, for throughout the year the temperatures of the emerging springs stayed fairly constant at around 55°F, while the spring and river waters were rich in dissolved nutrients acquired as the rain-water percolated through the chalk. Thus flooding not only raised the ground temperature, but it also nourished the pasture. First developed in the sixteenth century, the system had become widespread in the southern chalklands and south west by the seventeenth-century and expanded considerably thereafter.

In the steeper valleys, like those of Devon, Herefordshire and Oxfordshire, leats which followed the valley contours took water from a river or brook at an upstream point and thence along channels which were furnished with sluices. When a sluice was opened the water flowed slowly down to the parent stream, flooding the intervening pasture. A more simple system was used in parts of the Midlands and Wessex, where streams tended to have broad, level floodplains. 'Flooding upwards' simply involved damming a stream or river to pond-back waters and inundate the upstream section of the floodplain. 'Flooding downwards' was more complicated, but was popular in Hampshire and parts of Herefordshire. A watercourse was dammed and the impounded waters were distributed via a 'head main' onto series of artificial ridges or 'carriers', which had subsidiary channels running along their spines. A circulatory system was created, with the waters spilling from the carriers or 'drowners' into the intervening furrows, and back to the river. Such systems of drowning were operated regularly between December and March.

Drowning offered very considerable advantages, and Jane Doherty has described a farm at Itchen Abbas where, in 1801, the rent for water-meadows was fifty-five shillings per acre, while that for dry pasture was only twelve shillings. Even so, the mechanization of farming and the associated shedding of labour brought a virtual end to this useful but labour intensive method of farming, while the use of chemical fertilizers rather than the old muck and chalk stream nutrients has had unfortunate consequences for the diverse wild plant life which flourished under the drowning regime, despite the inundations and the heavy grazing which followed. One old meadow on the Itchen still contains over a hundred different plant species.

Though the old system of management has decayed, the characteristic channels and the ridged earthworks of 'floating downwards' may sometimes survive. The valleys of Hampshire chalk streams like the Avon, Itchen, Test and Meon were almost completely carpeted with water-meadows, and often the old carriers can still be recognized. Other continuous expanses of water-meadows can be seen by travellers on the Cambridge to Liverpool Street line, which follows the valleys of the Cam and Lee, while a particularly intricate system of channels in the meadows beside the Nar can be seen from the massive military earthworks of Castle Acre in Norfolk. However, one hesitates to recommend specific sites for fear that, between writing and publication, ploughing and reseeding or cropping will have obliterated the relics.

An idyllic scene of water-meadows beside the Taf near Whitland in Dyfed

CHAPTER 6
To Get to the Other Side: Bridges, Fords and Ferries

We have seen that rivers once played a vital role in the transport system and economy of Britain. But they could also act as barriers to cross-country movement, and the medieval period had long since ended before effective measures for the provision and maintenance of bridges were introduced. The ability actually to build bridges had existed for much longer. Indeed, the bridge experienced less in the way of architectural evolution than almost any other important construction. Medieval pack-horse bridges are not strikingly different from others which were being built at the start of the nineteenth century, and it was only the introduction of new materials – cast iron, steel and concrete – and the development of a whole new repertoire of engineering techniques during the Industrial Revolution which transformed the traditional outlines of the bridge.

The most distinctive and, apparently, most primitive type of bridge is the clapper, which is formed by placing massive slabs of stone horizontally to span the spaces between low, boulder-built piers. Doubtless the roots of this simple design burrow back into antiquity – but it is equally probable that existing clapper bridges are not particularly venerable. Most of these bridges span modest streams which are periodically swelled by floodwaters to become raging little torrents, so that the lifespan of any clapper bridge is likely to be terminated in a sudden and dramatic manner. Clapper bridges have a rather localized distribution and appear in areas where the traffic and resources have not merited the building of 'proper' arched structures, but where a supply of substantial slabs and boulders is to hand. There are a few examples in the northern uplands of England, more in the Cotswolds, while the best examples are found on the granite moors of the South West. One of the most attractive clapper bridges, quite primeval in appearance, is the Tarr Steps, crossing the Barle on Exmoor. More massive and imposing is the Post Bridge on Dartmoor, with four gigantic granite slabs carried on drystone piers. A few examples cross village streams, as at Linton in Yorkshire and Lower Slaughter in the Cotswolds. Less sophisticated even than the true clapper bridges are their little cousins, the clam types, with just a single stone slab to span the stream. Pedestrian crossings of another type may be provided by stepping stones. When found today they usually represent a recent and picturesque addition to the scene, as at Box Hill in Surrey, while the set at Studley Royal was a product of the eighteenth-century landscaping.

Most medieval bridges were timber constructions, and periodically they would be swept away by floods – to the anguish of all save the local carpenters. The whole topic of medieval bridges underlines the dilemmas of the society: the ability to build beautiful, durable and ingenious structures combined with the frequent inability to operate beyond the selfish little world of the estate or parish. Whether communities obtained the bridges that they needed greatly depended upon chance and the attitudes of local patrons. An urban community would gladly build a bridge if it could be seen that it would boost traffic and the custom at the

The bridge and bridge chapel of c1400 at St Ives in Cambridgeshire

markets, fair and workshops; the lord of a manor might similarly build a bridge if it would help to sustain the village market or improve the agricultural production of the surrounding estate, while abbots and bishops were generally keen to finance a bridge-building project which offered secure economic benefits. But without the existence of such patrons, corporations or charities, an area would depend upon ramshackle timber structures and hope that local benefactors could be found to foot the bill when the shaky edifice was swept away. In some places the switch to stone was finally prompted by the exhaustion of suitable sources of heavy constructional timber.

A fine bridge of about 1300 stands on older footings to span the Ouse between Huntingdon and Godmanchester, and it was probably financed jointly by the burgesses of the adjacent boroughs. It was typically well built in stone and carried the main trunk road until the recent completion of a Huntingdon bypass. However, had the common interest of the folk of Huntingdon and Godmanchester not existed it is likely that many centuries would have passed before the through route was properly served with a solid bridge. A few miles away is the superb bridge of around 1400 at St Ives. The river was probably originally bridged here in timber by the Abbot of Ramsey, around 1100, during his foundation of the town, and the subsequent provision of a stone bridge reflected the continuing interest of the Abbots in the prosperity of St Ives. The first St Ives bridge must have replaced a ford, located just upstream at the older village of Slepe.

This bridge is distinguished by the presence of a bridge chapel. Most important public buildings and places, such as bridges, markets and churchyards, were protected by crosses, although most of them were vandalized as 'popish symbols' at different stages of the Reformation and its aftermath. Bridge chapels were sometimes provided as places where travellers could give thanks for a journey safely accomplished, receive blessings for one about to begin and pay alms for the maintenance of the bridge. Other chapels grace the medieval bridges at Wakefield and Rotherham, while the more famous bridge at Bradford-on-Avon accommodated an oratory. Its roof was redesigned in the seventeenth century and it saw less exulted service as a lock-up and powder magazine.

Where, for one reason or another, a bridge was not provided then travellers would rely on fords to cross the lesser rivers and ferries to traverse the greater rivers and estuaries. Frequently the existence of an old ford is preserved in a village name. Perhaps the best example, and one which is often quoted, concerns the stretch of the Cam to the south of Cambridge. Here various branches of the prehistoric Icknield Way crossed the river at fords where, later, villages developed: Stapleford, Great and Little Shelford, Whittlesford, the separate settlements which merged and came to be known as Duxford, and the Chesterfords. Ford settlement names are usually much younger than the ford itself. Though Brentford includes the English '-ford' name element, this was the place where Julius Caesar described crossing the Thames by an existing ford, and the tribal patriots must

have known that he would use this crossing: 'There is only one place where the river can be forded . . . The bank had been fortified with sharp stakes fixed along it, and similar stakes had been driven into the river bed and were hidden beneath the water . . . . As the infantry crossed only their heads were above the water, but they pushed on with such speed and determination that both infantry and cavalry were able to attack together.'

With a few exceptions, such as the Thames at London, travellers crossed the broadest watercourses by ferries until quite recent times. For all their engineering genius the Romans depended on ferries to cross expanses of water like the Humber between Winteringham and Brough, and the Severn between Sea Mills and their great base by the Usk at Caerleon – as Christopher Taylor has described (*Roads and Tracks of Britain*, Dent, 1979). In medieval times ferries could reflect local entrepreneurship or charity, or else be demanded as part of the feudal obligations

The bridge and bridge chapel at Bradford-on-Avon

attached to the holding of particular lands.

The Mersey ferry, starring in the hit parade during the heyday of Merseybeat or the Liverpool sound in the 1960s, has a history extending back into the Middle Ages. In 1207 the burgesses of Liverpool obtained from King John the right to use the Birkenhead ferry free of tolls, though in 1316 the Prior of Birkenhead, whose monks operated the ferries, won the right to take a toll and payment from travellers using the Prior's accommodation. This was not the oldest Mersey ferry, and since 1178 the Knights Hospitallers, a crusading order devoted to the succouring of pilgrims, had provided a ferry between Runcorn and Widnes. In 1190 the yeoman, Richard de la More, was granted lands by the Hospitallers on condition that he and his heirs maintained a strong ferry boat. Rather than the service across the Mersey being grudging and unreliable, ferrying seems to have been a popular and lucrative occupation; the Prior's charges were increased at peak times, so that while the standard rate in the early fourteenth century was tuppence for a horseman and a farthing for a pedestrian, on Saturdays and market days the foot traveller paid a halfpenny, or one penny if he was carrying a heavy load. There were accusations of profiteering. Competing ferries must have proliferated, for in 1336 the Black Prince sought to regulate the number of operators. The tidal nature of the Mersey will have made it difficult to provide landings, and when Defoe crossed the two

---

Some Examples which Illustrate the Evolution of Bridges

Post Bridge, Devon, SX648788. A car park is available by the B3212 just SW of Postbridge village.

Moulton Pack-horse Bridge, Suffolk. The village is to the E of Newmarket on the B1085 between Chippenham and Dalham.

St Ives Bridge and Chapel, Cambridgeshire and *Bradford-on-Avon Bridge and Chapel*, Wiltshire, each lie close to the centres of their towns.

Teston Bridge, Kent. The bridge spans the Medway 3 miles SW of Maidstone, just S of the A26; a picnic area is provided nearby.

East Farleigh, Kent. The village and medieval bridge are 1½ miles SW of Maidstone, on the B2010.

Braybrooke, Northamptonshire. The village is on the minor road between Desborough and Market Harborough, about 8 miles SW of Corby.

The Monnow Bridge, Gwent. The fortified 13th century bridge spans the Monnow at the foot of the old high street in Monmouth.

Aberfeldy Bridge, Tayside. Wade's bridge lies between Aberfeldy, at the junction of the A827 and A826, and Weem, just to the NW of Aberfeldy.

Stackley Bridge, Cumbria, NY235109. A splendid pack-horse bridge on the Sty Head Pass track.

Ashness Bridge, Cumbria. Probably the most photographed pack-horse bridge in Britain on the narrow road from Derwentwater to Watendlath.

Ambleside Bridge and Bridge House (NT), Cumbria. The bridge is close to the centre of the tourist townlet, beside the main through road.

Bringewood Ford Bridge, Herefordshire and Worcestershire, SO454750. 3 miles W of Ludlow and reached via the side road from the A4113 which runs to Downton on the Rock and a footpath.

The Iron Bridge, Shropshire. 8 miles N of Bridgnorth on the B4373. Close by are the Coalbrookdale Iron Works, the Bedlam blast furnaces, the Hay inclined plane which lowered boats to the Severn, riverside warehouses and the Coalport pottery works, parts of a remarkable complex of industrial museums.

mile wide river at high tide at the start of the eighteenth century he described how passengers were carried from boat to shore '... on the shoulders of some honest Lancashire clown, who comes knee deep to the boat side, to truss you up, and then runs away with you, as nimbly as you desire to ride....'

The medieval traveller who had had his fill of ferries, fords and penny-pinching wooden bridges would have been delighted to approach a stone bridge, for where they existed they were usually well and skilfully built. Medieval bridges often reflected the evolving themes in Gothic architecture, with their pointed arches becoming flattened

A footbridge and ford on the Chess in the Chilterns

in the later examples, while post-medieval stone arches were usually rounded, in the Classical fashion. The typical medieval bridge was quite narrow, with recesses provided to allow pedestrians to avoid the traffic, pointed cutwaters projecting upstream to break the force of the water against the piers, and the undersides of the arches reinforced by stone ribbing. Pointed arches are prettily exemplified at Braybrooke in Northamptonshire, while ribbed arches are a feature of the delightful monastic bridge at that Mecca of river engineering, Fountains Abbey.

Scores of medieval bridges have been lost, some of them demolished and replaced during road-widening exercises, others during improvements to navigation. The Medway was noted for the magnificence of its medieval bridges, but several disappeared when the river was improved, deepened and partly canalized between Tonbridge and Yalding. The bridge at Rochester of 1387-92, built on the profits of French booty, must have been a robust and impressive structure, fourteen feet wide with eleven pointed arches. It was so vital to regional commerce that in 1449 the Archbishop of Canterbury offered a remission of sins to those contributing to its repair. In 1856 it was replaced by an iron bridge, itself later replaced. Of the surviving old Medway bridges, the best are at East Farleigh and at Teston.

Pack-horse bridges are sometimes regarded as a special category, but any public bridge of the Middle Ages, Elizabethan, Stuart and Georgian periods will have carried convoys of pack-horses. Equally, many of the lovely post-medieval stone pack-horse bridges of the Pennines and Lake District, which mainly date from the late seventeenth to early nineteenth centuries, will have been used by pack-horses, but were often built and used too by members of the local farming community. Perhaps the finest of the medieval pack-horse bridges is seen spanning the tiny Kennett at Moulton near Newmarket – a gem of a bridge which would surely have been replaced had Moulton grown or the traffic on the little track ever amounted to very much. One of the most spectacular of the much later northern pack-horse bridges is the one that carries the old road over Sty Head Pass down towards Borrowdale, hanging above a deep cleft of the stream. Perhaps the loveliest example is the renowned Ashness Bridge on the narrow Derwentwater to Watendlath track – but the strangest Lakeland bridge is surely the little one at Ambleside, which carries a house on its slender back. Perhaps built as a picturesque summerhouse of old Ambleside Hall in the seventeenth century, the diminutive bridge house certainly experienced domestic occupation, and housed a family with six children in the 1840s.

No discussion of medieval bridges can be complete without mentioning the exceptional Monnow Bridge at Monmouth, with its central tower gateway guarding the approaches to the town. At the castle town of Warkworth in Northumberland the bridge has a tower at one end.

The medieval forms of bridge design persisted into the middle of the seventeenth century, when modest changes influenced by the changing architectural trends were assimilated, with the adoption of wider, flatter rounded arches and lighter piers. A good example of a bridge of about 1640 can be seen at Clare College, Cambridge. A quickening of the commercial pulse and the dawning of the turnpike era of road improvements resulted in a great surge of bridge building and widening. Meanwhile, the English occupation of the Scottish Highlands produced metalled roads and stone bridges where none had previously existed. Under General Wade (1673-1748), who was sent to report on communications there in 1724, an important programme of works was launched, providing the rugged

A superb medieval bridge at Fountains Abbey; note the typical pointed and ribbed arches

The renowned and much photographed Ashness Bridge on the track to Watendlath; Derwentwater is seen in the distance

Grey heron. Its deceptively tranquil stance belies its
lightning speed when stirred into action

Nutrient-rich waters of the Waveney encourage rich bank flora in contrast to the cattle-trampled margin of the opposite bank where the flora changes. A typical traditional Broadland scene

A moorhen sitting on its nest

The moorhen is usually a shy creature, preferring the shelter of reedbeds

countryside with a basic network of military roads and bridges which were essential to the movement of troops and the pacification of the region by the army of occupation. Small streams were crossed by fords and culverts, but a number of good stone bridges were built and several survive. The most elaborate example is the Aberfeldy Bridge, designed by William Adam and built in 1733, with five arches, raised quoins and voussoirs and four obelisks in the elevated central span. It cost £4000, one hundred times the cost of one of the lesser bridges. More typical is the example, now bypassed by the Fort William to Kyle of Lochalsh road, which carries a stretch of military road over the Shiel close to the Glenshiel battle site.

Drastic changes in the design and construction of bridges were heralded by the opening, in 1781, of the Iron Bridge across the disaster-prone Severn gorge. Built by Abraham Darby III, this is arguably the most famous monument to the Industrial Revolution. The first Abraham Darby had leased a small charcoal blast furnace at nearby Coalbrookdale in 1708, and had pioneered the crucial technique of using coal in iron smelting by converting it to coke, and thus removing most impurities. By the time that the cast iron bridge was built, the Darby dynasty had developed Britain's largest iron-making business. The cast iron components of the bridge were jointed and bolted together in a manner reminiscent of carpentry techniques, and when they were assembled the bridge was two hundred feet long, forty-five feet high and with a central span of a hundred feet. The threat of collapse as a result of the convergence of the river-banks was recently met by reinforcing the channel with concrete. Another important monument to industrialization in the Welsh Marches is much less well known: Bringewood Forge by the Teme just west of Ludlow. The old seventeenth- and eighteenth-century buildings at this once

The fortified medieval bridge at Monmouth

Darby's Iron Bridge, opened 1781, over the Severn, perhaps the world's most famous historic bridge

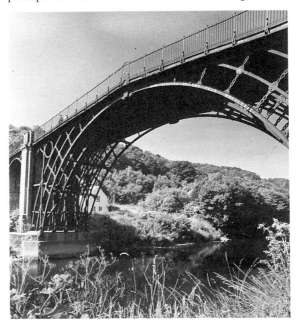

flourishing iron forging site have now almost disappeared, but a superb eighteenth-century bridge survives, with the overgrown wharf which served the iron works lying just downstream. The surroundings are clothed in the exotic flowers and shrubs of the neglected pleasure grounds which the ironmasters of the Knight family created at Downton Park.

The Iron Bridge achievement, the increasing momentum of the Industrial Revolution, the rising competence and ingenuity of British engineers and the demands of the railway age all helped to transform the traditional concepts and techniques of bridge building. A stone bridge of the late eighteenth century differed in no remarkable respects from one built by Romans – and gave little inkling of the shape of things to come. Soon there would be marvels like Thomas Telford's Conwy suspension bridge of 1826, with its castellated towers to harmonize with the great Edwardian castle nearby, or Brunel's daring Clifton suspension bridge of 1830-64, incorporating suspension chains bought

The lovely but litle-known bridge at Bringewood Forge near Ludlow

cheaply from his Hungerford bridge, demolished to make way for Charing Cross station. Robert Stephenson's High Level bridge at Newcastle was completed in 1849 in cast iron, carrying road and rail traffic across the Tyne on separate decks. The Forth Railway Bridge of 1882-90 was designed by Sir John Fowler and Sir Benjamin Becket. It stands 360 feet above high water, has a total length of 5300 feet, and a former world record main span of 1700 feet. But there were also a few disasters; Sir Thomas Bouch's Forth bridge plans were scrapped after the catastrophic collapse of his Tay bridge in 1879. The twentieth century has also had its triumphs and its problems, as expressed by the graceful Forth road bridge of 1964, the Severn Bridge of 1966 and the glorious Humber bridge of 1981, the two last named examples spanning stretches of estuary which had posed unanswerable challenges to the bridge builders of all previous ages.

CHAPTER 7

The Upland Stream

Brooks, cascades, rivers, marshes and estuaries are all manifestations of the eternal flow of water from high ground down through the valleys, across the plains and on to the sea. Associated with each stage in this journey is a range of different habitats. It includes the miniature worlds of the flanking rocks; stream-bed boulders, waterfalls, rock pools, beds of shingle and gravel, earth banks, muddy floors, islands, fringing swamps, and overhanging trees. Changes in water-level, caused by the lie of the land and fluctuations in the weather, add other dimensions, while today river valleys and plains are often largely managed by man, with man-made dams impounding water to create reservoirs and canals, aqueducts and drainage programmes all distorting the natural pattern. The result is a wealth of different wetland habitats, each with its own particular niches for wildlife. In this chapter we explore the little worlds of the upland stream.

The two essential requisites for life, oxygen and light, are more limited in water than in air. Oxygen is only slightly soluble in cold water, less so in warm water, and water creatures have to resolve this problem either by drawing fresh air supplies from the water's surface, or by developing a thin body wall, or instead, a specialized region which can extract oxygen from the surrounding water. Light, essential to photosynthesis in the plant world, cannot penetrate far into deep or cloudy water. However, there are few British rivers of any great depth, so that, unless water is made murky by suspended detritus, light usually reaches to the bottom. In fact, in running water the 'productive zone' is mostly restricted to the bottom, where shelter and anchorage are available. The currents and tides and variations in the volume of water flowing from season to season and from year to year may bring fresh supplies of oxygen and food and remove wastes, so that many organisms have been able to adapt, compete and survive. Therefore, in tracing rivers from their humble headwaters – perhaps no more than a moist spot in a forgotten pasture – to the eventual union with the sea, one encounters a kaleidoscope of all sorts of habitats and niches for plants and animals. In the upland section of the stream each river plant and animal has somehow solved the problem of how to survive in the mass of rushing water that lunges at everything in its path as it tumbles and gushes along its course.

Many British rivers flow through two different landscapes: upland and lowland. Some western situated rivers have no lowland section. These flow for only a few miles, plunging from steep hillsides via rapids and waterfalls, before tumbling into the sea. Equally, there are some lowland rivers, mainly in the south and east of England, which may have no upland section. Other rivers, being much longer, pass through a diversity of landscapes. The Severn, Britain's longest river, has its origins in the high acid rocks of central Wales, descends into the agricultural valleys of eastern Wales and through the Shropshire plain until it becomes a tidal river north of Gloucester. And so in Britain we find a series of rivers which range in character from the natural and unrestrained to

Some Upland Stream Habitats in the Scottish Highlands

Corrieshalloch Gorge (NTS), NH 204777. 12 miles SE of Ullapool. Mile long wooded gorge of R Abhainn Dorma includes Falls of Measach. Mixed woodland of native species; large variety of mosses and liverworts.

Torridon (NTS), NG 9059. Visitor centre at junction of A896 and Diabeg road in Torridon village. Wide range of stream habitats of mountainous country supporting a corresponding variety of plant and animal life. Where streams cross lime-rich rock and there is an absence of grazing, 'rock gardens' of alpine flora develop. Golden eagle and peregrine falcon live on high slopes. On lower ground fine bog communities develop, including spotted orchids, bog asphodel, sundew and butterwort. A few greenshank inhabit bogs. Red-throated diver, golden plover, wheatear, ring ouzel and meadow pipit are reported on lower slopes.

Cairngorms (NCC★), NJ 0101. SE of Aviemore, E of Kingussie. Approached via B970. Numerous streams dissect this granite plateau eventually draining into R Spey. Flora predominantly of acid rocks. Many rare mountain species, including arctic mouse-ear, alpine lady-fern and starwort mouse-ear. Extensive woods of native Scots pine – as at *Rothiemurchus* reserve★, visitor centre at *Inverdruie*, NH 903109, where access to the forest can be arranged – some at least 250 years old, accommodating a diversity of plants including the rare pinewood orchids, lesser twayblade and creeping lady's-tresses; insects, particularly a high population of wood ants which build their nests in pine needle mounds; birds and mammals. Roe deer roam the hills, valleys and woodlands, the squirrel is common in the mixed plantations and badger and otter tracks may be seen by the streams and lochs.

★Permit may be required. Contact authority for detail.

intensively managed watercourses which serve us as little more than drainage channels.

Upland watercourses are generally in their exuberant infancy, providing fairly unstable habitats. A trickling summer brook may in winter become turbulent or even a raging torrent, as is often the case in the highlands of Britain, when large amounts of water locked up as snow are suddenly released in the spring thaw. The permeability of the bedrock will affect the volume and speed of water that runs off, while the chemistry of the rock will affect its relative acidity and the associated nutrient and mineral availability for plants and animals.

The nutritional rating of fresh water is referred to as its 'trophic status'. The main categories of freshwater types are 'eutrophic' (alkaline waters) and 'oligotrophic' (acid waters). There are three further variations. At the two extremes, when alkalinity is greater than eutrophic, the rating 'marl' may be given, and when acidity is greater than oligotrophic, the water's status can become 'dystrophic'. When water shows both oligotrophic and eutrophic characteristics it is defined 'mesotrophic'. The continuously changing nature of rivers in the course of their journeys from source to sea will take them through most of the different trophic states, but where predominant rock types are uniform over large areas, water flowing through these areas will show specific characteristics.

The uplands of Wales, northern England, Scotland and the north and extreme west of Ireland are composed of tough rocks like slate, granite and gritstone, all of which impart acidity to the water: here streams tend to be oligotrophic. Over large expanses of highland areas glaciation has left depressions in deep permeable boulder clay, which is composed of rock dust and boulders dumped by ice (see Chapter I). In large flat areas where acid peat beds form and where rainfall is over eight inches a year, blanket bogs may develop. Such areas are mainly dystrophic and will support little flora and fauna.

The main plant types of such bogs are sphagnum mosses. Their numerous weak, slender shoots cluster to form close dense cushions which are wonderfully adapted to draw up water by capillarity from below, like the wicks of oil lamps, as well as holding rain-water falling from above. If one takes up a fresh wet tuft from the surface of a bog and squeezes it, the effect is similar to squeezing a bath sponge. The combination of permanent waterlogging of the ground, negligible oxygen supply and nutrient-poor water is ideal for both the growth of sphagnum and for the further accumulation of peat. As bacteria which

Devon

Dartmoor National Park. The R Bovey, Dart, Erme, Meavy, Teign which drain the moor are all torrential because of the heavy rainfall on these uplands. Their well-oxygenated waters support rich insect, fish and bird life. Vast areas of the park are heavily grazed moorland, heath and bog habitats typical of granite, but in the steep-sloped river valleys are stretches of rich woodland habitats. Many are nature reserves. The Dart Valley reserve (DNPA) is one of woodland and heath. The NT owns 2½ miles of *Holne Woods*, SX 700705, on the S bank of the river, upstream of New Bridge. On the banks of the river, oak grows with hazel, birch, holly and beech. The lack of air pollution encourages a splendid growth of mosses, ferns and lichens. An abundant aquatic invertebrate population favours salmon and trout. Dipper, wagtail, woodcock, green woodpecker, marsh tit, whinchat, ring ouzel, meadow pipit and skylark have all been recorded in the reserve and the woods provide suitable habitats for otter, mink, badger and fox.

Derbyshire

Longshaw Nature Walks (NT), SK 267800. On the outskirts of Sheffield on the edge of the Peak National Park. Various walks through a fine range of upland gritstone habitats, along stream and to bogs in open moorland and through a wooded river valley.

North Yorkshire

North York Moors National Park. The R Derwent, Dove, Esk and Rye have cut valleys into England's largest expanses of heather moorlands, mainly on acid sandstone. Where drainage is poor, bogs have developed, holding plants such as sphagnum, cotton grass, craneberry, bog rosemary and sedge species including the uncommon bog sedge. Bog bird species include curlew, dunlin and golden plover. The moors are managed for red grouse and therefore support predator bird species such as kestrel, merlin and peregrine. But about 150 resident bird species have been recorded in all the habitats in the park. Valleys enriched by limestone, which occurs in some parts of the park, support a richer range of flora and fauna. *Farndale*, SE 666974, at the headwaters of the Dove is famous for its spring daffodils.

break down oxygen cannot exist in these conditions of low oxygen, plant remains build up. Sphagnum further increases the acidity of the habitat as its lifestyle involves the exchange of nutrients for hydrogen ions. Few other plant species can cohabit with this plant other than specially tolerant species such as bog pond weed and some flowering plants, the best known of which are the insectivorous sundews. Others include bog bean, bladderworts and butterworts. Some species of the sedge family are also found in acid bogs, the beak-sedges being true bog plants. Animal life in this environment is minimal but for a small number of insect species, the occasional breeding pairs of migratory waterfowl and a few other bird species (mentioned below).

If a central trickle should develop conditions will become marginally more favourable, as with running water the supply of nutrients will be better spread. The movement of water will also ensure a better supply of oxygen. Many mountain streams are born of trickles from upland bog environments. They may be stained brown by acidic peat residues or brown algae. Rusty brown deposits are often the result of the presence of bacteria associated with the oxidization of ferrous compounds, a conversion process from which they will derive their energy. Such murky trickles may flow for but a few miles before meeting steep slopes with impermeable bedrock – and immediately the speed, volume and clarity of the water change as does the stream-bed. Local variations in the erosive power of the stream and the toughness of the bedrock can produce a succession of potholes, rapids and cascades.

Erosive action in upland limestone regions is downwards as much of the action of water takes place through solution of the rock (Chapter 2). Where water has been in contact with limestone or chalk it will have dissolved some of the lime and other minerals thus increasing in alkalinity. Such streams, eutrophic in nature, will create desirable habitats for many more plants and animals than can be found in upland oligotrophic streams.

An eroding upland stream is a turbulent habitat where survival depends on being able to hold fast. Where water cascades and flow is erratic, microscopic algae and diatoms cover boulders in a thin film, and a few lower plants such as liverworts and mosses like *Eurhynchium rusciforme* and *Fontinalis antipyretica* cling tightly to rocks and are able to survive submerged. In marginally slower sections where enough silt is deposited to obtain a foothold, conditions will favour only those higher

Water is oxygenated as it flows over rapids

plants which are flexible enough to yield to the currents. The fauna found in these waters is adapted to reduce water resistance, as we shall describe.

The freshwater habitat presents special problems for higher plants, due largely to

North Yorkshire

Yorkshire Dales National Park. Valleys of the R Aire, Ribble, Swale, Ure and Warfe carve their way through the park which lies mainly on Carboniferous rocks. Steep ravine slopes support clues to the original woodland, such as at *Ling Gill* (NCC★), SD 803778, a fine example of a limestone woodland habitat in a steep-sided gorge of a tributary of the Ribble.

Malham Tarn (NT and FSC: permit needed off rights of way), SD 890672. A wealth of plants and animals of bog, fen, woodland and grassland habitats. In spring, several species of saxifrage and other lime-loving plants are found under the scattered ash trees along the streamside path to the *Malham Cove*, SD 900270, while dipper and grey wagtail nest in the cracks and crevices of the spectacular grey-white cliffs.

★Permit may be required. Contact authority for details.

the fact that the essential oxygen is not very soluble in water. Hence many water and swamp plants have evolved special tissues which contain air, and which, in the case of plants which are only partly submerged, provide passages via which air can circulate from the aerial parts down to the submerged roots. Plants in fast running water have less of a respiration problem, for the amount of oxygen dissolved in the water as a result of the colder temperatures and bubbling turbulence is greater than in any other body of water. Their problem is how to survive and establish footholds in a mass of fluid that tugs at everything in its path. Plants have therefore become adapted to offer very little resistance to the current. Once they are anchored and rooted in the sediment, long green fronds with streamlined thread-like leaves drift with the currents; various forms of water crowfoot are good examples. Lacking in tough vascular tissue (as absorption of water and nutrients takes place over the whole of the plant's surface) their stems break off easily with sudden surges of water, perhaps to establish and propagate themselves downstream, but the roots remain to maintain the plant's presence.

Plants have also adapted their leaves to their particular position in the water between the surface and the stream-bed. Thread-like leaves are obviously more suited to an emergent life, while broad, flat, thin leaves at the water's surface are more efficient in the use of light for photosynthesis. A waxy upper surface resists wetting and drowning. Remarkably, the same plant may produce different leaves in different situations or conditions. The lesser spearwort, found at stream sides, will produce all at the same time: broad surface-floating leaves, sword-like submerged leaves and deeper spear-shaped emergent leaves. Many plants, like the water starwort with its unusual frond-like submergent leaves, produce additional broad surface leaves in the summer when, as the water is

Many rivers flow through two different landscapes: upland and lowland. In Newlands valley in Cumbria, there is a distinct contrast between the bareness of the higher, moorland section and the tree-fringed lower valley

calmer and there is less risk of damage, there is maximum benefit from the bright surface sunlight.

Reproduction in water plants is generally vegetative, as pollen is easily damaged by water. However, insect, wind and water pollination do take place. The water crowfoots produce nectar-sweet white flowers and pollination is aided by the insects attracted to them. Pond weeds produce spiked, petal-less flowers well above the surface of the water which are pollinated by the wind. Water starworts produce pollen containing water resistant oil which allows water pollination. This is important as the male flower has to drift down to a separate female flower situated lower on the plant stem before the stamen of the former can come in contact with the

Trickling mountain streams can become raging torrents. Llanberis Pass in Snowdonia

styles of the latter. Water birds aid dispersal, carrying seeds on their feet and feathers or through their digestive systems, or they may even carry small sections of plants which are established elsewhere and spread vegetatively.

As the water swirls and falls along its course the velocity of the current varies across the stream, generally decreasing towards the bottom. All brook creatures show great sensitivity in selecting their precise habitats within the stream. The most common animals are insect larvae which, by virtue of their greatly enlarged tufted gills, are well adapted to extract the dissolved oxygen, while their flattened bodies and clawed feet help them to cling beneath stones and between gravel particles. Streamlining is an important adaptation to enable easy movement through fast flowing water, allowing trips to the surface to

breathe. Mayfly nymphs, for example, have bodies which are flattened and which taper towards the rear to minimize water resistance. Free swimming animals may have powerful legs providing the strong propulsion needed to move in rapid currents. These legs are usually flattened, with long and hairy fringes increasing their surface area as they are thrust backward through the water, as in the case of many water beetles such as the whirligig and great water beetles. Other animals are furnished with suckers. On the surface of stones at the bottom of the bed there is a narrow layer of less than one-tenth of an inch known as the 'boundary layer' where the water is very slow moving. Here are found animals with flattened bodies such as flatworms and leeches. The former move relatively rapidly by means of the tiny hair-like 'cilia' on the underside of the body, but these can also serve as miniature anchors.

The leech has a flattened body with suckers at each end to hold the animal in place. The blackfly larva not only has suckers at either end of its body, but inside the walls of the larger posterior sucker are horny hooks ensuring the animal's anchorage even in the fastest flowing water. Hooks, grappling devices found on legs and other protuberances are all features used to hold on to the rock surface, but in other species still, the insect may first spin a silken mat on to the rock and then anchor itself to it. The blackfly larva uses a secretion from the salivary glands to spin a small silken mat on the stone surface, and using the hooks on its posterior it attaches itself to the mat. The larva then stands up and strains food from the plunging current through the free-swaying appendages on its head. Should it become detached from its mat and be swirled downstream, a silken thread attached from the mouth to the underlying stone or rock enables its 'proleg' to haul itself back along this silken dragline and reattach itself. Anchors and lines are also

Flowering plants, including the invading monkey flower and Himalayan balsam, have established themselves on the low banks of the Nidd, where nutrient-rich sediment is deposited when the river floods

The very rare fen violet survives at the National Trust
Wicken Fen reserve in Cambridgeshire

used by mature animals. Stone-flies may coat their eggs with an adhesive, while the eggs of some mayflies extrude long sticky threads that snare the stream-bed. However, weight alone is sometimes an effective anchor. Freshwater molluscs living in rapid flowing brooks and streams invariably have much harder and heavier shells than their relatives in lakes and ponds.

The art of encasing oneself to increase protection is the survival tactic of a few freshwater animals, and perhaps the most familiar, especially to anglers, is the caddis-fly. There are many kinds of caddis larva and most encase themselves in open-ended tubes made from bits of vegetation, sand grains, stones or shells built up on a silken envelope which the larva spins around itself. A pair of hooks attached to the last segment of the abdomen latches into the walls of the cylinder and this 'home' is moved along as the larva moves around. The design of the cases varies, and some of these constructions

Streams which pour down from the moors across acid rock in winter floods contain little life. But some plants and trees protected from grazing and exposure cling onto the sheltered faces of the gorges as here, above Haweswater, in Cumbria

are elongated into elaborate silken funnels which channel bits of organic flotsam into a central collection point. The edges of these shelters are often extended to form snares, securely anchored but able to billow out in the rush of water, trapping small insects and plant matter to supply a nourishing catch for the largely omnivorous caddis larvae. The larvae of *Limnophilus* will use materials such as small stones or snail-shells to weight their cases and avoid being swept away. Later they may discard the heavier material and make lighter cases of plant material. Inside the cases, the soft bodies of the larvae are fairly well protected against water currents and predatory insects, but fish and water birds will prey on the entire creature, case and all. *Anabolia nervosa* has, however, evolved a technique of attaching small protruding sticks to its case, which prevents it from being eaten – at least by some small fish. As well as protecting the soft bodies of the larvae, the cases also offer effective camouflage.

This wealth of invertebrate life feeding on algae, moss and other invertebrates is in turn food for typical fish species of upland watercourses such as brook lamprey, salmon and trout. The brook lamprey spends its entire life in brooks, streams and the upper reaches of rivers where there are sandy or gravelly beds. Spawning takes place between April and June. The eggs are laid in spawning troughs, after which the adults die. The eggs hatch in a few days and the larvae live on the stream-bed buried in the sand. They begin metamorphosis in the autumn an incredible five or six years after hatching, and the process is complete by the following spring, when they prepare to spawn.

Salmon migrate from the ocean to the river headwaters where they began life to spawn. Migration is thought to be triggered by an increase in the volume of water flow – a 'freshet' – that may occur during the spring thaw or after heavy rains, and two waves of fish come up the river, in spring and autumn. Spawning is initiated by a drop in the water temperature. The young can remain in fresh water for up to five years before making their way down to the sea. After spawning the female salmon

---

Scotland

The Tay. The tributaries of the R Tay have excellent stretches of upland stream habitats, although public access to some streams and rivers is limited as they have important salmon and trout fisheries. Some habitats can be examined in NTS property.

Ben Lawers (NTS and NCC★), NN 6138. Mountain and moorland habitats. Gravelly mountain springs and streams hold arctic-alpine communities, some of very local distribution. Rich insect fauna, including small mountain ringlet butterfly. Visitor centre, car park, nature trail.

Hermitage (NTS), NO 0142. Gorge and woodland reserve near Tay's confluence with R Brann. Dipper and wagtail nest near the cascade on the river and can be watched from a bridge over the gorge. Nature trail by arrangement.

Pass of Killiecrankie (NTS), NN 917627. 3 miles NW of Pitlochry. Oak woodland on the steep banks of R Garry. Rich ground flora including many local flowering plants such as bird's nest orchid. Waterside and woodland birds: dipper, grey wagtail, great spotted woodpecker, woodcock. Nature trails arranged from visitors centre at Pitlochry. Further downstream is the *R Tummel* nature trail (FC★), NN 865597. A walk leads through mixed deciduous woodland. Birds include redstart, wood warbler, spotted flycatcher, jay and greater spotted woodpecker. In 1953 a 50 foot high dam was built just below Pitlochry, creating *Loch Faskally*, NN 9358. Beside the dam has been built an enclosed 'fish pass' where salmon can be seen passing through ponds as they move upriver.

★Permit may be required. Contact authority for details.

---

immediately returns to the sea, but the male fish may remain to mate again before drifting downstream. At this stage many may die from disease and exhaustion, but those which do reach the sea can recover quickly, to return to the spawning grounds some eighteen months later.

It is still possible to see vast numbers of salmon migrating in some less developed upland Welsh rivers and the rivers and burns of the Scottish Highlands and Borders. However, in England, the increasing poaching and pollution of many rivers, particularly in their lower reaches, and the effects of damming without adequate provision of salmon ladders have greatly reduced salmon stocks. Conservation measures are sometimes taken. Restocking of rivers with young salmon from hatcheries and the provision of fish ladders around obstructions will, to some extent, remedy the damage and on the Thames the control of pollution has seen the return of salmon.

Trout are much more widespread than salmon: in the past few years there have been enormous developments in trout-farming to stock the reservoirs and the enclosed lakes. The brown trout is native to Britain, but the swift-growing introduced rainbow trout is generally preferred by the stocking authorities. In the uplands the Scottish and Welsh rivers are favoured trout environments and in England conditions in the Yorkshire Dales and the South West are also quite favourable. Much of its success derives from the fact that the trout is a very adaptable fish and can survive in the colder mountain streams of Scotland to the warmer waters of south-east England. In the more fertile swift flowing streams the lovely slate-grey and silver grayling often shares its environment with the trout. In contrast to the grayling the common eel is widespread, and will even tolerate situations uncongenial to most other fish.

Around the margins of the upland streams are insects, birds and mammals which may come to feed, drink and breed. Some will establish territories in the vicinity, depending almost entirely on the water habitat for survival, and others may just visit to drink or to seek occasional variety of diet. Insect breeding abounds in the stream habitat, as can be seen from the numerous insect larvae and nymphs which inhabit the waters. A most welcome sight is that of dragonflies, damselflies and butterflies. The authors have often caught sight of the female golden-ringed dragonfly hovering back and forth over the rippled surface of sheltered streams in the Lake District delicately extending her abdomen into the water to deposit eggs. Some species, such as blue aeshna and highland sympetrum, are confined in their range to Scotland and northern England. Terrestrial insects such as butterflies are restricted by the distribution of the food plants of their caterpillars. The large heath butterfly inhabits boggy areas in northern Britain where the white beak-sedge grows. The short-lived mountain ringlet usually occurs only above 1800 feet, where its caterpillar can feed on mat-grass. In limestone areas the rock rose provides food for the rare brown argus. Common species such as the small white, green-veined white and meadow brown may be seen even on dull days in summer where streams dissect woodland.

The nutrient-poor acid waters of many mountain and moor streams support fewer bird species than more fertile upland streams elsewhere. Curlews forage in streams entering or leaving lochs in the Scottish Highlands, while in the same country, near little rivulets flowing through bogs where bog bean provides good cover, the rare red-necked phalarope is reported to nest. Occasional pairs of mallard, teal and wigeon may nest beside streams on peat moss, feeding their young on whatever insect food they find. Red-throated divers select nesting sites in the safety of these secluded habitats, but may fly long distances to find food

The art of encasing itself to increase protection is the survival tactic of the caddis larva

The brown trout is very adaptable, being able to survive a range of waters, but upland rivers are favoured trout environments

elsewhere. Skylarks and meadow pipits, scattered far and wide over the uplands, may occasionally be seen by these streams.

In contrast, more characteristic of the more fertile sparkling waters of limestone areas are the dipper, the grey wagtail, the common sandpiper and the ring ouzel. Dippers, brown birds with snow-white bibs, invariably build their nests over water, to provide their nestlings with security from predators. Almost any support will do: a gap in the woodwork under a bridge, a natural ledge under a waterfall or bank or just the support of an overhanging tree branch. They are able to dive into the swiftest currents and descend to the bottom, where they 'walk' using their wings as flippers and search for aquatic insect food.

In summer, one of the loveliest of our small birds, the grey wagtail, can sometimes be spotted stooping to catch an emerging mayfly. Often it shares the stream side with the pied wagtail. Its habit of bobbing even more emphatically than the dipper and the wagtails makes the common sandpiper another cheerful feature of upland streams. Streamside drystone walls and rocky cliffs and ledges are favoured by the ring ouzel. Where the flow slackens red-breasted merganser may nest in the thick cover of the stream-banks or under larger well-lodged boulders.

Usually seen, if one is very lucky, as a flash of iridescent blue as it skims above the water, the kingfisher also nests in the steep banks of streams, where it excavates a tunnel more than half a yard long ending in a nest hole. It is found in England, Wales and Ireland, but is now only rarely seen in southern Scotland and is unknown further north. In 1963 the kingfisher population was decimated by a hard spell of frost that froze almost every stretch of fresh water throughout the British Isles, and the population is only gradually recovering. However, pollution of many rivers is eliminating the kingfisher from many areas.

In waterside oak-woods where there is a rich insect fauna, other species may visit or nest. These include the great spotted woodpecker and summer visitors such as whitethroat, redstart, pied flycatcher and wood warbler. In some Welsh valleys a scarce summer visitor, the pretty yellow wagtail, may be encountered.

The upland course of a river may also support a limited range of mammals. The water vole is a resident of the stream-banks, feeding on the shoots and underground stems of water plants, and the diminutive water shrew is sometimes observed crossing even the swiftest flowing rivers. Daubenton's bat, sometimes known as the water bat because it feeds over fresh water, may be seen patrolling the stream for insects, especially winter moths.

Where it is not hunted to near extinction, the otter is still to be seen, or heard, though more likely to be evident are its riverside tracks and slides left behind in soft mud. The otter holt is always well hidden within the roots of some waterside tree or in the more cavernous crevices of rocks and boulders where it is well protected from erosion. Several factors have contributed to its scarcity, including river pollution, mainly from poisonous farm chemicals and, shamefully, from hunting or trapping by water bailiffs along trout and salmon reaches – although otter hunting has been outlawed in England since 1978. Although it feeds mainly on eels, it is quite partial to fish of all types. In the uplands otters have now retreated to rivers in isolated hills and moorland.

Water can add an aesthetic touch to the most desolate of upland scenes, but by

The otter, now in danger of extinction because of pollution and the loss of suitable, secluded habitats

of other plant species may also be evident, particularly during the warmer months of the year. Colourful species of liverwort, lichen and moss may include the conspicuous red-tufted liverwort *Pleurozia purpurea*, the dense bright green moss *Philonotis fontana* and the silvery encrustation of the lichen *Cladonia impexa* – a vital source of food for various wild animals during winter. Some sphagnum mosses are also aesthetically pleasing in their appearance,

carrying food minerals and oxygen it can also transform a bleak habitat by encouraging a greater diversity of plant life, at least along the corridors through which it flows. Mountain trickles flowing over rocks rich in calcium offer good hunting grounds for some of the rarer arctic and arctic-alpine British flora such as alpine forget-me-not, spring gentian and purple saxifrage. Damp streamside ledges and slopes often form colourful natural gardens where plants such as roseroot, mountain sorrel, alpine lady's mantle as well as many species of saxifrage become established.

Where streams run down mountain slopes to flow through valley bogs below, a much richer flora than that of the blanket bog habitats, described earlier in the chapter, develops. Such waters will carry down the organic matter from the mountain vegetation as well as any sediments and minerals that have been eroded from the bedrock and now accumulate below. Sphagnum mosses will still cover the surface of these enriched bogs, but jewel-like spots

---

The Dee

One of Scotland's famous salmon and trout rivers, its banks and hillsides carry woods of great variety. One of the country's few large rivers unaffected by hydroelectric schemes and little polluted in its upland reaches. Access is limited.

Muir of Dinnet (NCC★), NO 4399. E of Dinnet at the junction of A93 and A97. Quiet waters and plenty of cover provide nursery area for otters. The reserve is famous for its landscape features formed by ice and water towards the close of the Ice Age among which are deeply cut melt-water channels – the Burn O'Vat occupying the deepest, along which there is a rich flora of mosses, lichen and ferns. Silver birch dominates the large area of woodland among which grow an abundant ground flora and interesting fungi, the late summer species including fly agaric, chanterelle and woolly milk-cap. The shallow *Lochs Davan* and *Kinrod* from which outlet streams flow into the Dee, display a wide variety of bog plants. 140 species of birds have been recorded in the reserve including resident and migratory wildfowl. More than 380 species of moth, including rare creatures such as the scarce prominent and large belted clearwing. Also many dragonflies, damselflies and beetles. Further upstream is *Dinnet Oak-wood* (NCC★), NO 468980. One of the few remaining oak-woods in NE Scotland with the character natural to upland oak-woods.

★Permit may be required. Contact authority for details.

such as the wine red hummocks of *Sphagnum magellanicum* or the orange-green leafed *S. recurvum*. Large hummocks of sphagnum may support cotton grass and shrubs of heather and cross-leaved heath. Flowering plants may include delicate trails of bog pimpernel and marsh-pennywort, the fragile white *Montia fontana* – which has been given the uninspiring common name 'Blinks' – the yellow-spiked bog asphodel, or even the bright rose-pink lousewort.

On the lower slopes the watercourse may dissect steep-sided wooded valleys which are very much features of the Lake District, many parts of Wales, the Pennines, western Scotland and Ireland and the upland plateaux of Exmoor and Dartmoor. On harder rock, characteristic of acid soils, sessile oak-woods are common, often found in close canopy but sometimes stunted and small. Woods that are on ground that is economically inaccessible for any form of timber production may even be fragmented relics of ancient woodland. Downy birch, rowan and holly may grow among the oaks. The shrub and ground layers are often poorer than those in lowland pedunculate oak-woods and, where grazing by sheep or other woodland animals occurs, they may be absent altogether. Otherwise, bramble and bilberry will develop. A notable feature of the ground layer in sheltered valleys is the abundance of ferns; hard fern, the buckler ferns and lemon-scented fern can all be found in amongst the spray-splashed rocks and boulders and mud deposits held in the large roots of the waterside trees. Creeping about amongst the mosses and liverworts in tangled masses may be found the unusual moss-like Tunbridge filmy fern. Flowering plants which may establish themselves in sheltered moist hollows include acid-tolerant species such as ivy-leaved bellflower, yellow pimpernel, wood sorrel and bluebells, while the yellow-flowering semi-parasitic cow-wheat thrives in oak-woods, deriving extra nutrient from host

---

Cumbria

Lake District National Park. The largest park in the country, rising from tidal shores to England's highest mountain: Scafell Pike. Most of the park's bedrock is acid but a narrow belt of Coniston limestone gives rich green river valleys in contrast to the bleak moor streams and bog habitats. Some fish found in Cumbrian waters are glacial relics such as char and whitefish, as is a small freshwater shrimp *Mysis relicta*, rarely found in Ennerdale. Many nature trails offer the opportunity to study some upland habitats.

Loughrigg Fell nature trail (NT), NY 375047. Leaves from Ambleside crossing the R Brathay to climb the slopes of the fell, demonstrating many features of upland and moorland river and stream habitats.

White Moss Common (NT and CNT), NY 348065. 2 miles NW of Ambleside on A591. Lake stream, woodland and open fell provide a picture of typical Lake District country.

Johnny's Wood Walk (NT and CNT), NY 254139. 8 miles S of Keswick above B5289 road to Seatoller. Passes over open fell and streamside oak and ash woods.

Nether Wasdale trail (CTNC), NY 147048. E of Gosforth, unclassified road off A595. The trail includes woodland, marsh and bog habitats in the R Irt valley, with dramatic views into the fells.

Borrowdale (NT), NY 245137. Several walks in the valley of the R Derwent, noted for its semi-natural woodland, especially rich in ferns, mosses, liverworts, lichens and spring flowers.

---

oak trees and woody plants, such as bilberry, in order to flower and set seed.

The First World War imposed an intolerable strain on Britain's woodlands. As a result the Forestry Commission was set up in 1919 to reafforest many of the devastated areas. It did so by introducing exotic conifers which have become familiar if unwelcome features of many hilly regions and often dominate much of the scenery of upland Britain and Ireland. About three-quarters of our commerical woodlands are formed of extensive tracts of serried lines of coniferous trees such as Sitka spruce, larch, or lodgepole pine. Conifers grow much faster than most hardwoods and therefore yield a return much more rapidly. However, they not only impart acidity to the water (see page 130) but also provide poor waterside niches for flora and fauna. Wildlife is deterred by the artificial nature of these dark, intensively managed, uniform habitats, and so food chains cannot develop.

Foresters are now more aware of the effects on the wildlife of the intensive nature of their operations. Some mixed plantings

A profusion of ferns and mosses grow on the rocks surrounding the sheltered gorge of this upland stream in Snowdonia

Wales

In Wales there are many beautiful areas with upland streams, such as in the Brecon Beacons and Snowdonia National Parks. Elsewhere other river valleys such as that of the Llugwy show features of an upland river, as do the streams at the Head of the Rhondda valley: *Blaenrhondda Waterfalls Walk*, SN 922021. Details from Mid-Glamorgan CC or FC.

BRECON BEACONS PARK: POWYS

Ystradfellte, SN 930135. 9 miles NW of Aberdare on unclassified road off A4059. Riverside walks along R Mellte through narrow limestone gorge with waterfall. Sessile oak-woods and rich variety of ferns, mosses and liverworts. Springtime brings many flowering plants before trees come into leaf.

Craig-y-Cilau National Nature Reserve (NCC★), SO 188159. 8 miles from Abergavenny on unclassified road off A4077, via Llangattock. A geologically diverse site rich in wildlife. Five rare species of whitebeam grow here on the limestone slopes, together with beech, yew, wych elm, oak and silver birch. Where tiny streams trickle over the scree limestone polypody and mossy saxifrage are seen among moss-shrouded boulders, while bogs are fringed with round-leaved sundew. Forty species of bird breed in the reserve.

SNOWDONIA NATIONAL PARK: GWYNEDD

Coedydd Maentwrog (NT and NCC★), SH 666417. Several entrances. Small car park off B4410 Rhyd to Maentwrog road. Steep oak-wooded slopes in the valley of Afon Dwyryd: one of the few large native oak-woods in Wales. Lichens, mosses and ferns are abundant. Spring flowers include celandines, primroses, violets, and butterflies, such as fritillaries and hairstreaks, may be seen. Many bird species.

Coedydd Aber National Nature Reserve (NCC★), SH 663720. Unclassified road off A55, S of Aber. The reserve includes oak-woods, streams and moorland dissected by the gorge of Afon Rhaeadr bearing the Aber Falls.

★Permit may be needed. Contact authority for details.

have been introduced to supersede the uniform and unbroken tracts of trees. In the Highlands surviving fragments of ancient and semi-natural pinewoods bearing our native conifer species – Scots pine, yew and juniper – are recognized as areas that need protection. Here shrub and ground layers will develop where there are breaches in the canopy, as along a watercourse, and a most interesting feature of these layers is the wealth of fungus species. As trees are different sizes and at different stages of growth and decay and dead wood is not cleared away, rotting material is available to provide the food and living conditions of species such as the shining brick-red capped sickener, or the conspicuous gelatinous orange witch's butter on decaying trunks and stumps. The easily recognized fly agaric with its white-spotted red cap is common in pinewoods in late summer and autumn, as is the pinewood mushroom, which grows amongst the needle litter in autumn.

By contrast, on calcareous soils over limestone, nestling in steep-sided gorges and ravines carved out by the eroding stream and its cascades or along the courses of 'gills' or 'cloughs', scattered woods of trees with a preference to lime appear, such as ash, common juniper, yew and hazel. Associated species also tolerant of acid soils are rowan or mountain ash, blackthorn, hawthorn, bird cherry, crab-apple and alder. The Carboniferous Limestone cliffs above the River Usk, near Crickhowell in Wales, are famous for harbouring rare whitebeams, and the lime-loving whitebeam appears in many local sub-species. The ground flora is rich, offering a bewildering variety of flowering plants, including rock rose, lily of the valley, baneberry, mountain pansy, melancholy thistle, wood and water avens, wood cranesberry, fairyflax, spotted orchids, herb robert, lords and ladies, foxgloves, valerian, violets and the insectivorous butterwort among others, and in a few special places the rare bird's eye

---

Durham

Upper Teesdale (NCC). Access strictly limited. An interpretative station, the *Bowlees Visitor Centre* (DCCT), NY 907283, is about 4 miles from Middleton beyond Newbiggin on B6277. The centre explains the natural history and land uses of Teesdale and displays some of the specialized plants rescued from the flooded valley (Cow Green Reservoir). There is a nature trail along streams, through woodland, upland turf and hawthorn scrub, to the fern-rich dampness of the spectacular High Force waterfall. Streamside birds include dipper and wagtail.

Hamsterley Forest (FC), NZ 093313. 7 miles NW of West Auckland, off A68 at Witton-le-Wear. Information centre, forest drive and nature trails. Largely planted woodland, but pockets of native trees – alder, ash, silver birch, grey and crack willow – grow along Bedburn Beck and its tributaries. The pattern of undergrowth reflects the soil types, the richest soils being beside the beck bearing bluebell, ramson, wood sorrel, common dog violet. The river was once wider than it is at present and so old meadows also lie in the river valley. Many woodland and river insects and birds, and mammals include roe deer, red squirrel, badger and fox.

---

primrose or dark red helleborine can be found.

On the lower slopes where the soils are deeper steep-sided valleys may be thickly wooded and further tree species appear such as gean or wild cherry, horse chestnut, sycamore, elm, elder and beech. Among the shrub and ground layers, bramble, snowberry and guelder rose are in evidence and woodland flowers include snowdrops, primrose, moschatel, woodruff and bluebells.

On the botanically famous upland plateau area of Upper Teesdale there are patches of 'sugar limestone', rock which has changed into its crumbly crystalline state after contact with very hot volcanic rocks. On the

a

b

Plants in riverine habitats can be grouped according to the zone they occupy. Among the flowering plants shown on pages 145–148 are some fully adapted for an aquatic existence while others are found in shallow water on the marshy fringes or on the drier ground of the river-banks:

a Yellow water-lily. Common in many aquatic habitats and difficult to eradicate once well established
b White water-lily. Much less common than the yellow-flowering species, restricted to only the cleanest waters
c Bur-reed occurs on the shallow margins of a river among reeds and rushes
d The yellow flag iris flourishes where reeds do not swamp it. Insect pollinated, the flower's dark brown honey guides on its petal attract the creatures

c

d

a Marsh-marigold, an early flowering plant, adds colour to river margins, before other species flower
b Butterbur. The male and female flowers are on separate plants
c Southern marsh orchid is a close relative of the early marsh orchid and probably its hybrid
d Water forget-me-not. Its creeping rhizome enables this plant to become well established in marshy meadows and along the banks of slow flowing streams
e Marsh-mallow. A brackish water species, found in the upper parts of salt-marshes, or as a 'casual' along marshy stream-banks

a Fritillary, a species found in marshy meadows, is now declining due to land drainage. Late summer grazing aids their establishment as, after dispersal, the seedlings make a better start in shorter grass

b The fragile ragged robin attracts the swallow-tail to its nectar in the Norfolk Broads, aiding cross-pollination

c The large flower-head of hemp-agrimony is attractive to butterflies, but it reproduces vegetatively

d Great willow-herb can grow up to six feet high, a striking feature if it grows in dense clumps

e Bee on purple loosestrife

f Yellow loosestrife occurs on fertile alkaline soils. It is not related to the more widespread purple loosestrife

a

b

c

a Water avens is common in the north but rare in the south. Where it grows near wood avens, hybridization into many different forms will take place

b Monkey flower, an introduced species that has become naturalized

c Himalayan balsam, another alien invader, has spread easily along stream margins, because its ripe seed pods will explode at the slightest touch scattering the seeds widely

d A rare species of upland limestone habitats; Teesdale violet, found on the 'sugar' limestone of Upper Teesdale in Yorkshire

d

Liverworts are common on river-banks and in damp shady places. The spores are produced on the underside of the strange umbrella-like rays

lime-rich soils here grow rare mountain plants, surviving remnants of a sub-arctic flora which flourished when the glaciers were retreating. They include spring gentian, Teesdale violet and Teesdale or bog sandwort, the last being unique to that area in Britain. A large expanse has been sacrificed to the Cow Green Reservoir but what remains is a National Nature Reserve.

Although many upland watercourses flow through sufficiently desolate areas to escape the consequences of farming, other human activities are having a dramatic effect on river life. Application of fertilizer to forestry land which is eventually washed into the river's waters is affecting the balance of natural plant and animal communities. Upland rivers frequently pass through grouse and sheep grazed moorland which is sometimes drained and regularly burnt to improve the heather and pasture. If this takes place in winter little damage will be done to breeding birds or flowering plants, but in summer everything including the dried up peaty soil will burn, endangering habitats.

Mining and quarrying discharge fine inorganic material into adjacent rivers and streams, which can clog up the gills of fish. River regulation and hydroelectric schemes are often placed in the headwaters of a river system and upland regions are favoured because rainfall is high and the deep valleys form natural reservoirs. Until the recently disclosed effects of acid rain pollution, these areas were also thought to be less affected by pollution, therefore putting a lighter load on purification plants. But the resulting irregular and massive discharges from reservoirs create a harsh and inhospitable environment in the course lying below the dam. Invertebrates are dislodged and plants desiccate or are submerged by the irregularity of the flow. Fish spawning has suffered from such interference. For example, the Llyn Brianne Reservoir completed in 1972 has destroyed salmon spawning grounds in the upper reaches of the Tywi river in Wales. Water discharged from the reservoir is excessively cold and has made a ten-mile section of the river sterile. Efforts have been made to trap the fish and transfer them upstream, but these have been unsuccessful. There is now much concern that unless the river is restocked artificially, the Tywi, once one of the finest salmon rivers, will have lost its fish by the end of the decade.

As the river flows on, conditions downstream may improve and life will reappear, but there is no escaping the fact that upland stream habitats and niches have been destroyed, possibly for ever, and waterways that were once rich in wildlife may now be nothing more than lifeless drainage channels. Thankfully there are still many places where one can follow the upper courses of a river through glorious areas of unspoilt countryside to explore its humble source.

CHAPTER 8

The Lowland River

Roughly south and east of the Tees-Exe line lie the lowlands of Britain. The highest areas are the hills of the South Downs and of Charnwood Forest and nowhere do they achieve a height of more than 820 feet. Broadly the soils are composed of clay vales and river deposits separated by escarpments of sandstone, chalk and limestone. From an ecological point of view the broad contrast between clays and sands and the chalklands is reflected in contrasts between characteristic flora and fauna: the chalkland is species-rich, whereas the clays and sands generally support a comparatively impoverished range of vegetation. This contrast is reflected in the streams and rivers and their floodplains. Those rivers which receive most of their waters from chalkland catchments such as the rivers of Hampshire – the Avon, the Test and the Itchen – are highly calcareous, supporting a diverse aquatic flora and fauna. In contrast, the river floodplains and the broad levels of Somerset, Sussex, Kent and the East Anglian counties tend to develop towards marsh and fen vegetation if not artificially drained.

The pattern of riverine habitats in these areas of Britain is closely bound up with the history of human land use. Where the land has been spared reclamation for agriculture, on chalk it supports woods or grasslands famed for their rich meadows. The undrained sand and clay floodplains have developed into wet woods dominated by alder and willows existing in more or less permanently waterlogged conditions, but where the fields serve as traditional pasture, botanically rich habitats have developed. In

north-east Norfolk there are forty or so man-made 'lakes' in the river valleys formed from enormous peat cuttings. The Broadland river fens of these famous Norfolk Broads are some of the finest examples of freshwater marshes in this country, as described in Chapter 3.

The speed of the current decreases as the river enters its lowland stage. It now starts depositing finely ground rock debris rather than eroding and carrying it. The stream-bed becomes more silty and the water becomes turbid with suspended organic matter, further enriched with fertilizer run-off or sewage if the river cuts across agricultural or urban land. As a result, levels of nitrates and phosphates will increase, as will water temperatures. With warmth comes a loss of dissolved oxygen, now replenished mostly through surface diffusion or, in a few places, where turbulence aerates the water. These sluggish waters support a different range of plants and animals than those more turbulent waters further upstream, and the trophic condition of the water will influence the variety and abundance of species. The rich supply of nutrients encourages a really abundant 'phytoplankton' (small aquatic plant organisms which live drifting in water): where water is sluggish and where excessive amounts of nutrients pollute the water, algae may grow so profusely that they use up most of the available dissolved oxygen. Fish and other plants and animals may die as a result. But when in balance, the profusion of phytoplankton supports a population of 'zooplankton' – animal plankton composed of various species of

Variations on a Theme

Britain has numerous lowland rivers and streams although each area has its own major river system: the Tay, Spey and Dee in Scotland, the Tyne and Tees further south, in England, the Severn in the west, the Ouse in the east and the most famous of them all, the Thames in the south. All will have sections which are of interest to the naturalist. However, access to the banks of some of the smaller rivers may be easier, and perhaps more rewarding, especially if they do not have much industrial development. Two different stretches of the same river may show different types of habitat, reflecting the local variations in substrate, flow, depth and volume.

Taf Fechan (Merthyr BC), Glamorgan. The Taf Fechan flows through a fascinating range of habitat, the wide variation due to changes in the underlying rock from acid Millstone Grit to rich alkaline limestone. The valley ranges from dry slope to a torrential river and superimposed on this is the effect of man, where limestone has been quarried and parts of the woodland have been cleared for fuel and grazing.

The upper reaches reflect the limestone richness in woodlands of Turkey oak, small-leaved lime, ash, bird cherry and an understorey of dogwood, field maple and guelder rose, while lower down on the open acid slopes are great spreads of bracken. Quarrying has left behind hummocked grassland, thick with hawthorn, wild thyme and common bird's-foot trefoil. In contrast to the steep valleys upstream, the river flows through marshy areas edged with greater tussock-sedge, water avens and water figwort in its lower reaches.

The rich animal life of the river also reflects the variations in habitat. Upstream the river is busy with insects such as stone-fly and caddis-fly, while dipper and pied wagtail pillage its waters. The woodland stretches contain birds such as nuthatch, treecreeper, green and greater spotted woodpecker, willow and wood warbler. In the marshy areas downstream are found snails and small, shrimp-like gammarids, while brown trout lie among the weeds. Here the kingfisher may be seen. A rich variety of dragonflies, damselflies, moths and butterflies breed in the reedbeds. The Merthyr Naturalists' Society provides information and details of access to the river stretches.

tiny invertebrates and crustacea such as insect larvae and pear-shaped cyclops, fairy shrimps and waterfleas, which will take on the role of converting the minute plant life into food for larger animals.

Although the current towards the centre of the river will be quite strong, the larger surface area of the bed increases friction, which considerably retards the speed of water along it. Hence fine sediment, falling out of suspension, builds up, offering habitats for animals that choose to spend part or all of their lives in the secure sheltered spots buried down in the rich organic material or in tiny crevices between and beneath the rocks and stones. Well over ninety per cent of the population of freshwater shrimp may be found in this microhabitat. The invertebrate is not particularly streamlined and can ill-afford to encounter strong currents, so it clings beneath the stones. Crevices are sought by predators; one such creature is the bullhead, a small fish which waits beneath the stones, rushing out to seize unsuspecting animals drifting by in the current.

Many scavengers crawl at the bottom of river-beds, and a now rarer specialist of fast running weed-free, stony, sparkling chalk streams is the crayfish. These creatures hide under the stones during the day and emerge to feed at night. They are omnivorous, eating insect larvae, mussels, snails, other small crayfish, fish eggs, and dead fish. The young will also eat the roots of water plants. The crayfish is a sensitive indicator of really pure water. One isolated incidence of pollution can wipe out the entire down-

Some Nature Trails and Country Parks Showing Lowland Stream and River Habitats

If there is a watercourse near you it might be worth getting in touch with the offices of the District or County Council and the local Naturalists' Trust to find out if nature trails can be arranged. Information about the following may be obtained from the relevant authority.

Tregassick Nature Walk (CNT), Cornwall, SW 857340. S of Truro, off A3078. A trail following the banks of the tidal R Percuil.

Cotehele Nature Trail (NT), Cornwall, SX 423681. N of Plymouth, near Calstock. Trail through woodland and marshland habitats along R Tamar.

Frome Valley Nature Trail (BNS), Avon and Somerset, ST 622765. Riverside trail on the outskirts of Bristol where water and waterside flora and fauna can be observed.

Five Pond Wood Trail (NT), Avon and Somerset, ST 223321. SW of Bridgwater, near Broomfield. The trail runs through a woodland along a tributary of the Parrett.

Auckland Park Nature Trail (Wear Valley DC), NZ 215290. On the outskirts of Bishop Auckland. Trail in parkland around a tributary of the R Wear.

Barnes Park Nature Trail (Sunderland DC), NZ 383557. Trail on the outskirts of Sunderland along a tributary of R Wear.

Cathedral Peninsula Nature Trail (DCCT), NZ 272422. Urban trail along R Wear, Durham.

Styal Country Park (NT), Cheshire, SJ 835830. A park on the outskirts of Manchester through which R Dean flows; rich in both woodland and river life.

Elvaston Castle Country Park (Derbyshire CC), SK 413332. SE of Derby off A6. Woodland and streamside trail in the Derwent valley. Birds include mallard, tufted duck, kingfisher, grey wagtail and woodland species such as nuthatch and lesser spotted woodpecker.

Loggerheads Country Park (Clwyd CC), SJ 198626. SW of Mold off A494. Fast flowing lime-rich waters of Afon Alun are spangled with flowers of common waterfoot and a rich variety of bank species. Trout- and minnow-rich waters where there is an abundance of insect larvae and freshwater shrimp. Many river and woodland birds.

Ewloe Castle Nature Trail (Clwyd CC), SJ 292670. NW of Ewloe off A55. Trail through open farmland, woodland and along stream of Afon Alun.

Crathes (NTS), Grampian, NO 7389. E of Banchory off A93. Nature Trail in the ground of Crathes Castle along R Dee.

Drum (NTS), Grampian, NJ 7900. E of Banchory off A93. The old Forest of Drum has been a woodland site for several centuries and still carries magnificent mature beech, oak and Scots pine. Wych elm, juniper and ancient yews and wild cherries are also present. Nature trail along R Dee.

Newbold Comyn Country Park (Warwicks DC), SP 329659. Royal Leamington Spa Riverside habitats of woodland, grassland and marsh. Literature from Tourist Information Office.

Sherwood Forest Country Park (Notts CC), SK 6267. N of Nottingham, A614–B6034, just N of Edwinstowe. Car parks and information centre. Country park at the heart of ancient Sherwood Forest including magnificent oak woodlands through which flows R Maun.

Coe Fen Nature Trail (Cambridge City CC), TL 448575. Causeway S of Cambridge city centre. Land used for grazing for centuries. Nature trail runs through meadow-land along R Granta. Details from Cambient.

Clare Castle Country Park (Suffolk CC), TL 774454. Near Sudbury. Nature trail in the castle grounds in the lovely R Stour valley.

Downs Bank (NT), Staffordshire, SJ 901365. N of Stone. Heath, stream, marsh and woodland trail in R Trent valley.

(Elsewhere in this book the gazetteer features many riverside sites of interest where there is public access to the banks of the adjacent river giving opportunities to study the riverine wildlife of the stretch.)

A river loses momentum and begins to deposit material as it enters its lowland reaches. The Tamar, on the Cornwall-Devon border, near Cotehele

stream population of crayfish, and the river will be devoid of the creatures thereafter.

Also inhabiting the rivers' waters are a large population of molluscs. These include snails and limpets, which form one group, and mussels and cockles, which form the other. The main difference is the form of their shell. In snails and limpets the shell is all in one piece, while in the other group it consists of two parts hinged together. The shell is not merely an external shelter but an integral part of the mollusc, attached to the body by powerful muscles. The principal constituent of the shell is calcium carbonate and so waters seriously deficient in this chemical will generally show an absence of molluscs. The creatures move and anchor themselves by means of a muscular 'foot'

development on the underside of the body, and in snails the feature forms a flat 'sole' on which the animals can be seen gliding along the surface of mud or water plants. The nature of this foot also enables them to travel upside-down along the water's surface film.

The gently flowing, warmer, food-rich waters of the lowland river habitats not only support far higher insect populations than appear in the upland watercourses but at this stage in the river's course many other species appear. They can be grouped into aquatic and terrestrial species, and in some cases the adult forms are terrestrial and the larval stages aquatic. Mayflies, stone-flies, dragonflies, caddis-flies, alder-flies, lacewing flies and midges all spend their larval stages in the mud and among the stones on the stream-bed. The larva of the sponge-fly is a parasite of the freshwater pond sponge, which is usually found in the deeper parts of slow moving rivers. Our two common freshwater sponges are peculiarly

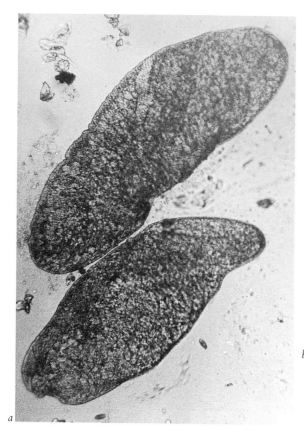

a

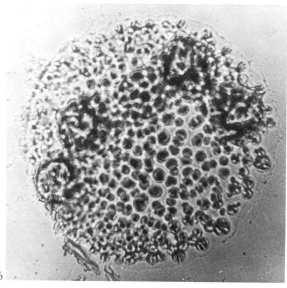

b

Freshwater micro-organisms: (a) *Spirostomum ambiguum;* (b) *Volvox.*

misnamed, as the river sponge occurs in lakes and ponds.

Unique among British lepidoptera, china mark moths also spend their immature stages under water. They prefer very slow moving waters and marshland as the female lays her eggs on the underside of the leaves of floating plants such as bur-reed or frogbit. Young caterpillars will gradually work their way into the tissues of the host plant, in this way obtaining food and shelter while breathing in water by means of thread-like gills. Eventually these creatures will form a protective oval case by joining pieces of leaf tissue with a silken secretion, and within this case the tiny caterpillar can remain completely dry although the structure is totally submerged. The larvae hibernate in winter when the vegetation dies down, only to become active again with the return of the food supply in spring, and shortly after they will pupate, either above or below the water surface depending on the species. Adult moths are on the wing from June to September and spend the day resting close to water, to become active at dusk.

Other conspicuous groups of water insects abound in slow moving sluggish canals and in the numerous pools, ditches and small streams which lie in the river corridors. The pools may be little 'cut-offs', the remains of old meanders, while the ditches are usually man-made drains in marshes, fens and farmland. Here are seen water-striders skimming across the surface, apparently never breaking the surface to get wet. They are voracious scavengers, mainly feeding on algae, but taking any opportunity to make a meal of various luckless insects that fall into the water and drown. The water-boatman swims just under the surface of the water, propelling itself by its powerful hind legs. Also very active and aggressive, it may foolishly attack creatures larger than itself, even fish – only to become their prey!

Such bugs, which spend much of their time swimming just beneath or on the water surface, are extraordinarily well adapted. Air breathers have water-repellent features to keep them from drowning. The narrow bodies of pond-striders and the smaller water gnat are kept dry by an upper waxy coating, while underneath, the air-trapping, dense hydrofuge hairs keep the insects upright. Water-repellent 'feet' are a response to surface tension (produced by a strong attraction between the molecules of water), so that each leg creates a dimple in the water surface, thus cushioning the creature as it 'walks' on water. Diving and water beetles display a highly efficient swimming mechanism, having evolved folding leg 'paddles' composed of blades and bristles. Upon the outstretched power stroke, these paddles fan open to provide maximum thrust, but with the recovery stroke they fold to offer little resistance to water, the complete cycle of action pushing the creatures swiftly through the water.

Sluggish water margins in the summer months are one of the best places to see various species of delicate and beautifully coloured dragonflies and their slender cousins, the damselflies. Most of these insects are ephemeral creatures of high summer, able to live in their adult stage for only a few weeks, though their larval stage may last many years. Both dragonflies and damselflies have similar life histories and commonly share the same flying, feeding and mating grounds. Both are aggressive predators, capturing insects with their legs while in full flight or swooping to snatch them off riverside vegetation. Damselflies and smaller dragonflies will often fall victim to the strong legs and powerful jaws of much larger members of the same family, such as the emperor dragonfly.

Male dragonflies are aggressively territorial, defending territories sometimes covering long stretches of water. Intrusion by other male dragonflies may lead to aerial battles and males are sometimes seen with pieces torn from their wings. These insects are usually doomed as they rely entirely on their wings for movement. Their legs have become so specialized that they can only perch on or cling to vegetation; they are unable to walk on a horizontal surface being further handicapped by the length of body extending behind the legs.

If a female visits the male's territory she will be courted, and on accepting his advances she will allow him to clasp her at the back of her head and together they will

Freshwater shrimp live under stones in clear fast flowing rivers and streams

A water-strider can skim and jump across the water's surface, able to keep itself upright because of the dense pile of air-trapping hairs on its underside

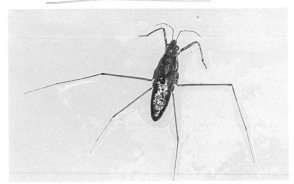

fly around in tandem to seek a suitable mating place. While she is thus clasped, other males are prevented from making contact with her. During mating the female bends her slim agile abdomen right back until she makes contact with the sperm capsule on the male's abdominal segment and her eggs are fertilized. Egg-laying varies from species to species, and for some this may be the final fatal act. For instance in the case of the emerald damselfly, also known as the green lestes, after fertilization the male retains his hold on the female and slides down with her along a reed blade to almost a foot beneath the water surface where she first makes a slit in the plant tissue with her saw-like ovipositor and then lays her eggs within. The mating pair must be particularly careful at this stage, and will drown if their wings become waterlogged.

After hatching, the length of time spent in the water by the nymphs depends on the species, and may even be up to five years. Nymphs are predatory creatures, capturing their food with an extension of the lower lip which shoots out to stake the prey on a pair of hooks. Towards the end of the larval stage of its lifecycle, usually in early summer, the nymph will leave the water, climbing out on the stem of a water plant. Its skin will slowly split down the back of the thorax, releasing first the head, then the thorax and finally the legs of the adult insect. After a brief resting spell the wings expand as air and liquid are pumped into them. Slowly the wings dry and the insect can then take flight, to live a month or so, long enough to breed, unless it becomes prey, before coming to the end of its life.

The emperor dragonfly, mainly a southern species, is one of our largest and most handsome insects, with a wingspan of over four inches. Their characteristic hawk-like habit of restlessly patrolling their territories gives hawker dragonflies their name. They include the common aeshna, and the brown aeshna, which is easily spotted, for its large yellow wings attract attention. Sturdier bodied darters are less restless movers, and many species are confined to the warmer lowland rivers of England. They include the common sympetrum, the black sympetrum, the keeled orthetrum, the brilliant emerald and the broad-bodied libellula.

One of our biggest damselflies is the banded demoiselle, found along fast flowing streams mostly in southern England, but also in parts of northern England and Wales. Other widespread species include the common ischnura, the common blue, the common coenagrion and the large red.

A search among the plants and leaves of waterside trees will lead to a discovery of the grazing caterpillars of several species of moth and butterfly. During the long summer days adult insects too can frequently be seen on the wing or at rest. Moth species include one of the most striking of British hawk moths, the eyed hawk-moth. Its caterpillars feed mainly on various kinds of willow and sallow. A smaller cousin, the elephant hawk-moth, lays its eggs on the great willow-herb and marsh-bedstraw. Other riverside moths include the attractive red underwing and the handsome herald, whose caterpillars also feed on willow; the emperor, which in spring may be seen flying strongly over water plants, its caterpillars sometimes abundant on meadow-sweet; the reed moth, whose caterpillars live on various water plants including reed sweet-grass; the rarer rosy marsh moth, and the water ermine, whose black-spotted, dark reddish-brown hairy caterpillars may be found on yellow loosestrife, water mint and water dock. Notable among the crepuscular species (adults active at twilight) are some of the 'wainscots' particularly associated with reed swamp and fen food plants and the dingy footman, whose caterpillars feed on the lichen found on alder and willow trees.

Among the butterflies, the rare swallow-

A water-boatman is adapted for buoyancy and so its trips underwater have to be made by clinging to plants

tail is probably the most colourful and lordliest. Unfortunately, it is now confined to a few areas in Britain where a watery fen environment is maintained to allow milk parsley to flourish and where the insect's larvae and pupae have been undisturbed, as at Hickling Broad in Norfolk. This umbelliferous plant is the caterpillar's chief food plant. It is sometimes also found on the now also rare ragged robin. The butterfly was once widespread throughout the Fens of East Anglia and possibly also in the marshes of the Thames and Lea rivers before land drainage, burning, mowing and the use of herbicides caused the decline of milk parsley. Caterpillars and chrysalises also fall victim to spiders, birds and small mammals, and like other insects the butterfly suffers from natural disasters due in the main to unfavourable weather conditions. In 1975 an attempt was made to reintroduce the swallow-tail at the Wicken Sedge Fen in Cambridgeshire where it had become extinct since the early 1950s. Some 3500 food plants were grown and 228 butterflies released, but the drought of 1976 followed and the butterflies and their offspring perished. The swallow-tail is now a protected species by the Wildlife and Countryside Act of 1981.

Another lovely butterfly, the large copper, also became extinct a century ago, when its disappearance followed the draining, mowing and burning of its home grounds. Today it may be found at Woodwalton Fen in Cambridgeshire, where it has been reintroduced with some success. Its caterpillars feed on great water dock. More common waterside butterflies include the marsh fritillary, which is attracted to devil's-bit scabious, the brimstone, whose caterpillars are found on alder buckthorn, the peacock, which is partial to the nectar of hemp-agrimony and meadow-sweet, and the small tortoiseshell, which is attracted to the strongly scented water-mint flowers.

One of the typical sounds of summer is the chirping of grasshoppers, and waterside species include the common field grasshopper which is one of the most widespread British species. More damp-loving species include the meadow grasshopper, the lesser marsh grasshopper and the now rare large marsh grasshopper, which seeks wet patches amongst bog asphodel and bog myrtle.

The water-spider, Britain's only entirely aquatic spider, hunts, courts, mates and moults under the water surface. It constructs an underwater retreat with silk secretions and the tissue of aquatic plants which it fills with air bubbles trapped in its body hairs. *Pirata piscatorius*, although not entirely aquatic, is also found almost exclusively by the water's edge. It roams about in the wet vegetation and when alarmed will run down the stem of a water plant into the water. Many terrestrial species may also choose to live in this insect-rich habitat and two familiar species which may be found on the flower-heads of reeds are the jumping spider and the crab spider. Both species are conspicuous on reed heads because the female often constructs her egg cocoon in such places. Young spiderlings do not crawl

The common blue damselfly is the most widespread of the small blue damselflies

Swallow-tail chrysalis. The adult will emerge from this overwintering structure towards the end of May, or in June

down from these high nurseries when they hatch out but will attach themselves to individual strands of silk to become airborne and float away when these threads are caught by air movements. This mechanism ensures their dispersal over a wider area.

This myriad of invertebrate life provides food for amphibians, fish, birds and mammals, the first two groups in turn providing meals for the last two groups higher up in the riverine food chain. The fen and lowland river marshes provide suitable habitats for the grass snake, which is particularly abundant in some of our southern fens, the common toad, the frogs and the newts.

Certainly far less common today than it was a few years ago, the common frog is found throughout Britain, and also in parts of Ireland, as it is an explosive breeder in early spring. However, a whole colony can

be wiped out in a season as a result of pollution, and the frog is disappearing in many places where agricultural chemicals can arrest the growth of the tadpoles, where breeding places are destroyed and where there is a depletion due to collectors. They are the natural prey of some birds and mammals, and herons are known to kill more frogs than they can eat at the spawning sites, leaving numbers of mutilated victims dying by the side of the water. The edible frog, which was introduced to Norfolk from the continent, was once very prolific, but is now almost extinct. The attractive marsh frog made a spectacular invasion early this century after being introduced into Romney Marsh in Kent, but it too is declining.

The three species of newt, the common, palmate and great crested, are also suffering from the dangers of pollution and a severe

Newly emerged swallow-tail drying its wings

loss of breeding places. Although the major part of their lives is spent on land, the creatures must return to water to avoid excessive water loss, as too much exposure results in death, and to lay their eggs. Ponds in gardens and nature reserves help to increase or maintain their numbers in a few specially favoured localities.

The common toad is widely distributed, but noticeably scarcer than it was in many localities only a few years ago. It is less dependent on water when it is not breeding, and may wander far to drier ground immediately after spawning. Massacres on roads from cars are familiar where migration routes cross the highways and the cumulative effect of the losses inflicted over several years can result in the extinction of a whole breeding colony.

More people are drawn to the banks of our rivers by angling than by any other activity and the growing popularity of what is said to be Britain's most popular participation sport is beneficial in so far as it is in the interests of both anglers and licensing authorities to keep the waters as pure as possible. But there are also some destructive side-effects. Not only has plant life on the river-banks been destroyed by the trampling of hordes of fishermen or controlled to prevent the snagging of lines, and animals and birds have died painfully as a result of carelessly discarded nylon line, hooks and lead weights, but also, to improve the sport, foreign species such as rainbow trout and the voracious zander have been introduced at the expense of some native species which have been culled, devoured or displaced.

About 5000 of the 20,000 known species of fish live in fresh waters, and of these only thirty-eight species occur in British waters. This relatively small number of native species is a consequence of the Ice Ages, which made most British rivers uninhabitable by fish. The rivers which hold the largest numbers of different breeds of fish are generally those in the east, south and south west, and those holding more than ten species include the Thames, the Hampshire Avon, the Exe, the Parrett, the Severn, the Warwickshire Avon and the Dee. For the purposes of examination of fish distribution rivers can be divided into biological regions from their headwaters to their estuaries and each region will very generally have its characteristic fish. The headwater stretch of rivers, usually characterized by stony or gravelly stream-beds and strong currents, is termed 'troutbeck'. Further along the course, where the current is still fast but some silting occurs in sheltered spots, the watercourse enters its 'minnow reach' zone. In the 'lowland reach' the current is slow and the course meandering. Here much silting takes place. Finally the river reaches the sea and the conditions and tidal action of an estuarine environment limit the fish to those which tolerate a high degree of salinity (see

Chapter 9). A typical sample river would thus carry the ubiquitous eel and trout almost throughout its length, but other species may be distributed as follows: salmon would ascend to spawn in the troutbeck; grayling, dace, chub, bleak and gudgeon would inhabit fast running water in the minnow and lowland reaches; pike, roach, perch, bream, tench, carp and rudd would be found in the lowland reaches only; and finally, river lampreys, flounders, bass and other saline tolerant fish would be found in the estuaries. Of course, due to a variety of integrating factors 'typical' rivers are non-existent, but such a method of division is a useful simplification.

The distribution of fish can also be examined on a geographical basis, while certain species are selective about the depth at which they feed: some fish are surface feeders and others prefer to lurk and feed at the bottom in deeper murkier waters as we shall see.

Dace, chub, bleak and gudgeon, all preferring clear running water, are found throughout most of England outside Cornwall. Bleak is almost absent in the Lake District. Dace, chub and bleak are all surface feeders, and on hot summer days can sometimes be seen near the surface or leaping out of the water after flies. Gudgeon are bottom feeders, living on shrimps and larvae of mayflies, caddis-flies and midges.

As the river becomes wider and slower, roach, rudd, perch and pike appear. A member of the carp family, the roach lacks teeth, so the size of food which it can take is limited. It tends to feed on insects and insect larvae, but if there is a shortage of animal food, it will eat vegetable matter instead. Its adaptable lifestyle and tolerance of a wide variation in temperatures enables it to inhabit a large range of habitats.

The rudd is often confused with the roach, and is possibly the most attractive of Britain's coarse fish, noted for its beautiful silvery sheen and red fins. It is a nervous creature, extremely difficult to approach as the shoals bolt at the slightest sound. Promiscuous fish of this species are noted for hybridizing with other closely related fish such as roach, bleak, bream and white bream.

Perch, a strikingly striped, humpbacked fish, is widespread and common throughout Britain and Ireland except for parts of the Scottish Highlands. The bigger they get, the more predatory they become, but few waters are capable of growing really big perch, and if there is a shortage of food the fish will eat their own young. Large fish become solitary creatures and are therefore difficult to catch, and we are told that when they are hooked they fight with surprising power.

Perhaps the most easily recognized and popular freshwater sport fish is the voracious pike. It is generally absent from the rivers of Scotland, West Wales and Cornwall, preferring slow flowing waters and good vegetation cover, where it lies motionless in wait for its prey. It is carnivorous at every stage of its lifecycle, and the adults take whatever is available, eating all sorts of fish as well as frogs, newts, water birds and aquatic mammals. They can live up to seventeen years by which time they can have grown to three feet in length.

When the rivers reach their flat lowlands and where meanders, ox-bow lakes and other backwaters develop, fat and lethargic carp, bream and tench appear. Tench are able to live in waters which have a low oxygen content and prefer to lead inactive lives down in weed-infested areas. Animal life at the bottom serves as their food, and in conditions of drought they are able to become dormant in the bottom mud.

The carps – crucian, common, mirror, wild and leather (no scales) – are highly regarded among anglers for their fighting qualities. Experienced fishermen claim that these fish have the capacity to learn and remember, as once having been hooked on a

The grass snake is an excellent swimmer and may hunt for insect larvae, tadpoles and fish in the water

Amorous common frogs. Frogs returning to their breeding sites often find them filled in or polluted and this, among other things, is leading to their decline

particular bait, if they get away they will avoid it ever after. The ability to evade capture and the fact that they are able to exploit several different sources of food enable carp to achieve mammoth proportions in rich waters. The fish have no teeth in the mouth, but instead have two bones on the floor of the throat thus swallowing their food whole to chew it in the throat. Carp have been valued for rearing since medieval times in thousands of

fish ponds and continue to be raised in fish farms today to stock angling waters.

The brownish common bream frequents slow still waters where it is usually solitary, shoaling only in the breeding period. As this fish grows, its diet changes until it becomes exclusively a bottom feeder, standing on its head to 'suck' food items from the bottom, its mouth parts being especially adapted for this purpose. The silver bream, identified by its silver scales, is much more local and a native of the Yorkshire and East Anglian rivers that flow into the North Sea. The ruffe is another small fish confined to East Anglian rivers.

In recent years different fish populations have been severely affected by diseases, like the one which drastically reduced perch in the early 1960s. Fish, like humans, have their own immune systems, but fish become vulnerable when stress lowers their natural immunity. Often this stress is man-induced and can result from artificial changes in water-levels associated with dams and sluices or from the traumas associated with angling. Current research is aiming to produce a better understanding of fish diseases.

Birds as well as anglers take fish. The bird life of riverine habitats includes several groups. Some true aquatic species depend entirely on the water for food, breeding and sanctuary. Other species will spend the winter in riverside marshes as visitors. Some birds found in this environment are not at all dependent on it and simply visit to add variety to their diet. However, those species which do colonize riverine habitats show their adaptation to life in them.

The handsome great crested grebe and the more widespread but less obtrusive little grebe (or dabchick) are both completely aquatic species. They feed mainly on small fish and large invertebrates but, expert divers as they are, they need a sufficient depth of water in which to gather enough speed to catch their food. They also prefer to

Shelduck up-ending

nest in the safety of deep water and build their nests from the stems and leaves of underwater plants. From a distance these floating platforms appear fragile and flimsy, as they project only a few feet above the water, but in fact the bulk of the nest is actually underwater and is firmly anchored to vegetation, usually reeds or rushes, but sometimes to the branches of overhanging willows trailing in the water. These floating nests are able to rise and fall with changes in water-level, but are liable to flood due to heavy wave action such as that caused by power boats. Another hazard is a sharp drop in water-level due to drainage and, less often, to drought, leaving the nest high and dry and vulnerable to prowling predators.

A delightful sight in both species is their courtship ritual. During the breeding season the handsome colours of the cock (the great crested grebe is far more spectacular than its smaller cousin) are best displayed. In the mating dance the partners approach each other 'standing up' breast to breast in the water offering each other strands of waterweed. Great crested grebes are particularly attractive when their appealing striped chicks are being carried on the parental back.

One is less likely to see little grebes in courtship because they are shy creatures, taking to cover at the slightest disturbance. But as sound plays an important part in their mating ritual, it is more likely that their shrill duet is heard than the birds themselves are seen.

As one would expect, many different ducks are numbered amongst the aquatic species, mainly feeding and breeding within the confines of fresh water. Two diving ducks which are often seen together in mixed flocks are the tufted duck and the pochard, but the two species do not occupy the same ecological niche. The tufted duck eats mainly animal food, particularly small molluscs and crustacea, while the pochard depends more heavily on the soft leaves and stems of underwater plants such as stoneworts. However, when feeding on plants the pochard will often take in the molluscs and any other invertebrates clinging to them, while the tufted duck will invariably swallow bits of plant while attempting to eat an attached snail. Both species nest in water among reeds and sedges near the shore or on islands in the middle of larger rivers.

There is a whole group of ducks called 'dabbling ducks'. Their feeding technique is that of dabbling on the surface or at the water's edge, moving along with their bills just inserted into the water, sifting small invertebrates and seeds floating near the surface by drawing water in at the front of the bill and squirting it out through the sides. Dabbling ducks include mallard, gadwall, teal, pintail and shoveller. The shoveller is a most specialized surface feeder, its large spatulate bill being able to increase the amount of water sucked in, while the

Greylag geese grazing on riverside vegetation. A few native birds breed in Scotland, but elsewhere they have been introduced

fine hair-like serrations inside both the upper and lower mandibles work as extremely efficient filters, trapping even such small particles as plankton. Groups of shovellers often feed communally, circling in close packs. Each bird filters in the wake of the one in front and as this feeding raft is in action it has the effect of stirring up the water, bringing more potential food to the surface.

Up-ending – tilting the body through 90 degrees – is another feeding technique practised by dabbling ducks, enabling them to feed on the bottom of river-beds. Various species of bird sharing the same area will feed without competing at different depths of water. The teal, being the smallest, is restricted to the shallowest parts of the river, while swans, having much longer necks, can up-end to reach down to much greater depths. The whooper and mute swans can both reach down to about a yard.

Apart from the aquatic species, there are many birds which depend on the river habitat for food and choose to take up waterside nesting sites. The kingfisher and dipper will often make their territories along the lower reaches of fast flowing rivers. The greater spotted woodpecker is known to destroy many of the goat-moth's wood-boring caterpillars in willow trees, while the lesser spotted woodpecker may even take up residence in rotten alders, and the nightingale is said to breed abundantly in the sheltered river valleys of Suffolk. Family parties of whitethroat, a common little summer resident, may be seen hunting for insects amongst the waterside vegetation, while the greenfinch is frequently seen on river-banks in autumn and winter. Swifts, swallows and house martins may nest some distance away from water but find the richness of the food supply attractive. They may sometimes be seen swooping over the water to take insects in flight. Sand martins live closer to the riverine restaurant, nesting in holes in exposed river-banks.

The unmistakable heron will nest in trees near water, or in reedbeds and there are heronries in all areas of the British Isles. Their feeding technique is fascinating to watch. The bird will wade slowly through the shallows of the river until it spots a suitable prey – eel, fish, frogs, newts and larger invertebrates – or, alternatively, stand motionless until something comes swimming or crawling past. When the victim has been spotted, the heron will be seen cautiously to stretch down its neck towards the spot, then make a lightning stab into the water, emerging victorious with the prey in its spear-like bill. Herons, like many other fish-eating birds are vulnerable to hard winters, when the rivers may freeze over, but populations have also been reduced because their eggs are thought to be affected by organochlorine pesticides. Scientists at the Monks Wood Experimental Station in Cambridgeshire claim that the eggshell of British herons is now only four-fifths as thick as forty years ago, and the eggs are therefore vulnerable to being crushed by the brooding parent bird.

Other waders feeding at the water's edge, probing for invertebrates in the mud and amongst the stems of water plants, include lapwing and golden plover, while on shingle banks the greenshank, green sandpiper and common sandpiper may also be seen searching for insects.

Marshy reedbeds along river corridors, as in the Fens and the Broads, are some of the most sought after habitats by river bird life. Here is found a variety of live food and many nesting sites as well as much shelter in the cover of the dense vegetation. Of the different species of birds recorded in Europe, about 190 are associated with fresh water, and most of these are found in marshland. Some of these species have become so adapted to living in a marsh environment that they are unable to adapt to other situations and have simply disappeared with the destruction of this habitat – the marsh harrier, bittern and bearded tit are often such victims.

The bittern, victim of marsh and fen drainage, is slowly becoming re-established in some large RSPB protected reedbeds of Norfolk. It feeds in the same way as does the heron, to which it is related, although it has a shorter thicker neck. Its mottled brown plumage is a perfect camouflage in the marsh vegetation, where it may stand statue-like, its beak poised ready to spear its prey. Hidden among the densest clump of reeds the creature is rarely if ever seen, but its booming call is unmistakable.

The sedge warbler arrives from Africa in spring to build a waterside nest from a mixture of plants, reeds and grasses. Other species which weave nests among the reeds include the marsh warbler, the reed warbler, whose nests are popular with cuckoos looking for foster parents for their eggs, and another close relative, the reed bunting which, unlike the warblers, does not seem very adept at nest-building, creating most untidy structures from dry grass, lined with softer and finer plant material.

The rare water-rail is another bird which spends a great part of its life concealed amongst the dense patches of reed. It is a shy sensitive bird which is most active at night, and like the bittern is more likely to be heard – a most unusual grunting and screaming call – than seen. The hen is so sensitive to disturbance that she will sometimes build two nests. If she feels her eggs are in danger in one nest, she will carry them one at a time to her other one. The RSPB reserve near Fowlmere in Cambridgeshire has a small population of these birds.

The coot and moorhen show their adaptation to life in the marsh as their awkward long legs end in widespread toes, to spread their weight when walking over boggy areas. The moorhen is one of the most common of our waterfowl, and although it is at home in the water, some birds are quite happy nesting away from it, instead taking to the trees. However, if it nests in water it will choose a site offering good cover amongst the waterside vegetation. Usually a shy bird, it will submerge itself when alarmed, if not near the cover of vegetation, leaving only its beak above water.

The coot will always build its nest in the form of a substantial platform in deep water, but usually surrounded by dense vegetation cover. These structures will be well anchored to prevent them from floating away in a flood. Although these nests will not rise with the level of water in flood conditions as do grebes' nests, the parent coot can add to their height within a matter of hours, taking the eggs up the construction out of harm's way.

Marshy areas provide suitable habitats for many species of duck and goose, both resident or those simply visiting to roost and feed. Teal, pintail and mallard will usually roost during the day and leave the marsh to feed in open water at night. Goose species include the white-fronted, pink-footed and less common bean, Canada and Egyptian, the latter two of which have escaped from reserves to establish breeding colonies in the wild, as in parts of Norfolk. The greylag goose, Britain's only native breeding goose, will roost in the marsh during the night and feed on the surrounding farmland during the day. It will pick at the wasted spilled grain or feed on the tiny potatoes and broken pieces remaining on the surface after the harvest, thus doing the farmer a good turn by reducing the number of rogue plants that would otherwise come up the following year.

Stretches of many lowland rivers are now maintained as nature reserves by the RSPB and Wildfowl Trust and these sanctuaries invariably attract an enormous variety of birds, both residential and migratory, in large mixed flocks which may not otherwise be able to survive and compete together. Under such protection we are fortunate to see the return, particularly in the fens and marshes, of scarce birds such as the Bewick's swan, the marsh harrier (a population of sixteen or seventeen pairs in 1977 is as

River Wildlife Sanctuaries Open to the Public

(Contact relevant authorities for visiting times)

Otter Trust Reserve, Norfolk, TM 315884. SW of Bungay off A143. Otter and wildfowl collection on R Waveney.

Peakirk Wildfowl Refuge (WT), Northamptonshire, TF 168068. S of Spalding off B1443. Wildfowl collection set on the site of an old osier bed and the Car Dyke. The site is visited by winter wildfowl and passage waders. Tributary of R Welland.

Martin Mere (WT), Lancashire, SD 428145. Burscough, N of Ormskirk. Wildfowl collection but the man-made mere and marshland area by the Leeds and Liverpool Canal attracts a spectacular range of wild birds. In winter the meadows provide a roost and feeding ground for many thousands of waterfowl. Over 10,000 pink-footed geese migrate annually from Greenland and Iceland to the reserve. Now becoming an important wintering ground for whooper and Bewick's swan.

Arundel (WT), Sussex, TQ 020081. NE of Chichester, off A284. Wildfowl refuge on R Arun where many wild and passage birds overwinter and breed in lakes, reedbeds and marshy fields, including bittern, bearded tit and Cetti's warbler.

Washington (WT), Durham, NZ 330565. SW of Sunderland, off A1231. Waterfowl park and refuge in pond, reedbed and woodland habitats on R Wear. Includes many species that are now declining in the wild.

healthy a one as we have enjoyed for many years), the bittern and the bearded tit. So eagerly sought after were the pretty and delicate eggs and nest of the latter, which bred in fairly vulnerable reedbeds, that the birds soon became rare.

Unlike birds, the mammals of riverine habitats are not as easily seen, as many are nocturnal and those that are active during the day seldom come into the open. Water is of vital importance to some aquatic species, but other mammals, though common in riverine habitats, are not exclusively found here. The otter lives closely associated with water and is well adapted, with its lithe body, webbed feet and muscular tail, for life in an aquatic habitat. It is aggressively protective of its territory, which may extend to several miles along a stretch of river, ensuring it gets its large daily requirement of food without too much competition. Fish constitute the bulk of the otter's diet, but when supplies are short they will turn to other sources of food including frogs, crayfish, molluscs, earthworms, water birds and a variety of small mammals. The creatures are even aggressive towards each other in courtship. A bitch will attract a dog otter to her territory by giving off a scent, but will then do her best to deter him. She will eventually accept him, but soon be rid of him after mating, to bring the young up on her own. The cubs, blind at birth, are suckled by the bitch until they are capable of accepting fish in their diet. Gradually the proportion of fish will increase, and at about three months the cubs will accompany their mother on fishing expeditions. The family unit will usually stay together until the cubs become fully independent at about a year old.

Otters may once have had a preference for the rich lowland river marshes of England and Wales, but hunting, insecticide pollution and the increasing disturbances of recreation have led to their serious decline in the last fifty years. The insecticide dieldrin,

used in agriculture and industry, infiltrated the freshwater systems and found its way up the food chains in progressively higher concentrations to affect fish-eating birds and animals. This, together with severe changes in habitat as a result of drainage schemes and for the convenience of recreation, seriously affected this secretive mammal, which seldom tolerates the presence of man. Thanks to writers such as Hugh Miles, Gavin Maxwell and Henry Williamson, public sympathy and understanding for this lovely and personable animal have been aroused in recent years. In 1971 the Otter Trust was founded under the directorship of Philip Wayre to concern itself with the worldwide conservation of otters. As a result, since January 1978 otters were added to the schedule of endangered species and are protected in England and Wales. The mammals are now being bred under protection near Bungay in Suffolk, and successful breeding in captivity is offering hope that they may return to some of their old haunts. After careful research into the creatures' social organization and dispersal mechanisms, trials are being initiated to release selected young otters, which are untamed and unattracted to man, into specially chosen habitats. The requirements of the chosen habitats are isolated stretches of river offering suitable shelter and a reasonable stock of fish, a high quality of water little affected by pollution and, very important, sympathetic landowners who are prepared to co-operate in the project and consent to the animals being released on their properties.

Another expert swimmer and diver is the water vole, sometimes mistakenly called the water rat, because it superficially resembles the brown rat. Although these creatures are often seen in or near slow moving water, they are also known to visit or inhabit places well away from aquatic environments. Their food consists predominantly of the stems and leaves of succulent grasses, and

Otters at the Otter Trust at Bungay, in Suffolk

they are partial to the tender roots and shoots of willow, but molluscs and insect larvae also feature in their diet. Their nests of grass and rush stems may be a cosily lined short burrow in amongst the roots of a waterside tree or even above ground in an abandoned bird's nest. Although the young are born naked and blind, they are virtually self-supporting at three weeks and some will be breeding in the year in which they are born. The breeding season is about seven months long, from April to October, and females may rear three to four litters in a season. Apart from being vulnerable to all manner of predators – including pike, heron, weasel, stoat and owl – voles tend to have a short life-span, and so such rapid breeding is necessary to ensure the survival of the species. Younger voles may take over the territories of older animals, forcing the latter to leave, but family groups may join forces to do battle against a predator. The males queue up in combat, attacking one at a time, one after another, so that each renewed effort and strength may have the effect of wearing out the intruder. Sometimes this battle force is of no avail, and many of the army will become the prey of the hungry predator.

A smaller cousin, the chestnut-coloured bank vole, although not aquatic may also choose to make its quarters in the drier sections of the river-banks. This creature is mainly vegetarian, and buds, berries, seeds and bark mainly constitute its diet. It can sometimes be seen climbing, precariously, to the tips of twigs overhanging water after wild fruits. Population explosions will also occur in this species which has similar breeding habits to that of the water vole, but large numbers do not usually survive for long because populations of predators, particularly short-eared owls, increase proportionately. When the floods of 1938 and 1953 greatly reduced the population of small mammals in thousands of acres of broadland marshes, it was very noticeable that birds of prey suffered in consequence.

Two aliens, both escapees from fur farms, have now established themselves along many rivers and streams in East Anglia: the North American mink and the South American coypu. The mink, a vicious killer, whose diet is composed of water voles, fish, birds and insects, used to compete with the otter for food and breeding sites. These aggressive animals always kill more prey than they can eat, so they are a menace to small mammals and nesting birds.

Unlike the mink, the coypu feeds extensively off waterside plants and root crops, and burrows into river-banks, often destabilizing the banks as a result. They have multiplied explosively, spreading rapidly throughout the fenland waterways, which have provided them with inexhaustible supplies of food in the form of aquatic and marsh vegetation. Breeding goes on throughout the year and female coypu can produce two or three litters a year with five to ten offspring in each. The balance of nature has been upset severely where mink and coypu have established themselves, for they have no natural predators, but both are now under attack by man.

From spring to autumn bats such as the pipistrelle or the long-eared bat will leave their cavernous hibernating sites to roost in riverside trees. At twilight they will begin patrolling back and forth above a stretch of river, hunting for insect food, meticulously discarding the insects' wings, which are sometimes found in leaf-like patches along the river-banks.

Steep, wooded river-banks may offer refuge to badgers, and in suitable areas a colony may establish itself. Their 'setts' may have many entrances, leading underground to a series of tunnels, and can be identified

A large and destructive animal, the coypu

by large heaps of excavated earth mixed with traces of discarded bedding – hay or bracken – at wide entrance holes. Spiders favour discarded tunnels, and an entrance full of leaves and cobwebs indicates no occupant. A certain sign, during the day, of animals asleep somewhere in their underground chambers is the presence of flies near the entrance. And look out too for badger paths and 'slides' in high river-banks nearby which, if muddy, will bear their tracks and footprints. These nocturnal animals usually emerge at dusk to seek food – fruits and berries, a variety of invertebrates (particularly earthworms) and even the young of small rodents – or a change of bedding.

Other mammals which may find a niche in riverine habitats include the red fox, which will find a plentiful supply of food among the waterfowl, small mammals and insects. Drier areas of secluded reedbeds and marshy carr may even provide suitable nursery areas for a vixen's cub.

Brown hares, known for their ability to swim, inhabit grazing marshes in the Fens. Here they are able to make their 'form' – a surface depression in the vegetation where their young are born. An extraordinary sight with which many country people are no doubt familiar is that of two hares standing upright facing each other in a boxing posture.

The helpless fledglings of birds nesting in reedbeds are particularly vulnerable to stoats and weasels and these fierce little carnivores find a rich and varied diet of amphibians, small mammals and fish in this environment.

This variety of insect and animal life, of prey and predator, as well as the many vegetarians, lives in association with the plants growing in the water and may find shelter among those growing along the banks, which include a profusion of flowering plants. Here we describe some of the 'typical' plant species associated with a particular river regime, but along the length of any single river local variations in depth, in the composition of soil and nutrient, and in management will occur, resulting in different assemblages of species.

In fast running, clear streams and rivers such as the Test and the Itchen in Hampshire and the Dove in Derbyshire, fed by the underground springs that emanate from the chalk and limestone areas of Britain, submerged plants flourish. If weed cutting and dredging are not maintained then these can form thick clumps covering the stream-bed and growing to the surface. There are numerous species of starwort, crowfoot and pondweed. Once established, Canadian pondweed will spread rapidly since any fragment that breaks off soon roots and grows into a new plant. It flowers only rarely and sets seed even less often. Instead, small buds forming in the axils of the leaves during the summer swell with food reserves and then drop off in the autumn and winter. Since each bud can grow into a new plant, its colonization can be very dramatic. Other submergent species include lesser water parsnip and watercresses.

The stable flow in chalk streams (see Chapter 2) ensures a high water-table for the species on the banks throughout the year. Undisturbed banks can be dominated by

Some Sites Showing Chalk River and Stream Habitats

(Contact relevant authority for details)

Titchfield Haven (Hampshire CC), SU 540040. On the outskirts of Fareham, near Titchfield village. Permission required from Naturalist Ranger in Fareham. Marshes and grazing meadows of R Meon. Important overwintering site because of its closeness to the Solent.

Lower Test Valley (HIOWNT), Hampshire, SU 364150. Access on public footpaths crossing the reserve, but special permit required elsewhere. 450 species of flora recorded. Many insect and bird species including winter wildfowl and waders.

Upper Hamble Country Park, Hampshire, SU 490114. SE of Southampton, approached via A3051. Riverside woodland, rich flora and fauna.

Oughton Head Common (Hertfordshire CC), TL 172307. On the NW outskirts of Hitchen a footpath leads to the banks of the R Oughton, a pure fast flowing chalk river. In the upper reaches a rich marshland has developed, abundant in insect fauna, particularly dragonflies and damselflies. In deeper clear waters downstream, ten-spined stickleback, bullhead, minnow and trout thrive.

Cuckmere Haven (East Sussex CC), TV 519995. W of Eastbourne off A259. R Cuckmere meanders through a wide meadow valley to end in a chalkland estuary providing a superb complex of habitats. The *Seven Sisters Country Park* lies in the reserve. Visitors centre in the park publishes a good trail guide.

Woods Mill (STNC), Sussex, TQ 218137. NW of Brighton between A283 and A2037. Wetland area in the R Adur valley, where lime-rich waters drain from the South Downs. Details from information centre in the reserve.

Fowlmere (RSPB), Cambridgeshire, TL 407462. S of Cambridge, off B1368. Chalk springs and derelict watercress beds in R Cam valley. In spring and summer there are many marshland birds, the principal nesting species being reed warbler and water-rail.

plants which require damp conditions rather than those which just tolerate them. Sedges and reeds are typical, and the greater tussock and lesser pond sedges are frequently found growing alongside reed sweet-grass and whorl grass which encroach into the water. Also on the banks are found purple loosestrife, one of our loveliest native plants, yellow loosestrife, hemp-agrimony, common comfrey, great willow-herb, blue water speedwell, water figwort, skullcap, water dropwort, globeflower and the monkey flower, a native of western North America, an alien so called because if the flower is held upside-down it is supposed to resemble a grinning monkey. Its red-spotted yellow flowers, conspicuous in July and August, produce a capsule containing around 150 seeds, which are readily dispersed by water or on the feet of water birds. As a result the plant is now widespread and still spreading. Policeman's helmet, or Himalayan balsam, a native of the Himalayas, is another plant which has spread almost everywhere along stream margins by means of explosively released seeds. A tall and attractive plant, its flowers in different shades of pink produce splashes of colour, although the plant is displacing native species on scores of river- banks.

Along the edges of chalky streams are often found the remnants of once widespread marshy woodland, though agricultural development has now confined the marshland trees to the edges of streams where alders and willow still flourish. Willows come in many shapes and sizes, local varieties being perpetuated by their facility to grow from broken shoots. Man for over 2000 years has used the willow to grow withies which are still harvested annually in some places, including sites near Burrow Mump in the Somerset Levels, to make baskets. Riverside willows are still regularly 'beheaded' or 'pollarded' to control their growth and so open up the river-banks for fishermen.

The plants of lowland streams which flow sluggishly over clay, silt or peat contrast markedly with those in chalk streams. The narrower, upper channels are often choked by branched bur-reed, while downstream, submerged and floating plants dominate. Free floating plants such as duckweed, its tiny, round lime-green leaves forming extensive mats, sometimes give sheltered bays or ditches a complete green covering. Rootless plants such as frogbit and the rare and threatened water soldier may also become established, their nutrients being taken directly from the water by fine rootlets hanging down beneath their leaves.

Yellow water-lily is common in slow rivers, but the white species is restricted to only the cleanest of waters. They can be seen flowering in close proximity at the National Trust reserve at Wicken Fen in Cambridgeshire.

Several species of pondweed, including the fennel-leaved and the large shining, also live happily in slow moving watercourses; the latter grows prolifically in the rivers of south-east England, but some of the grass-leaved species have declined in recent years because of pollution and habitat destruction.

The Somerset Levels and the East Anglian and Lincolnshire Fens display the significant features of fen vegetation. Along the many miles of the river valleys is a mixture of open reed swamp, sedge beds, herb-rich fens and alder woodland. Vast areas have been drained and turned to agricultural use since the seventeenth century, but some important fragments survive, many of which have been turned into nature reserves. Among the various examples are Wicken Fen and Hickling and Ranworth Broads in the Norfolk Broads.

One of the significant features of fen or marsh vegetation is its tendency to change gradually over a long period, due to the accumulation of layers of peat or silt. As the surface grows above normal water-level, the fen becomes drier. Aquatic plants such as

Some Sites Showing Broadland, River Fen and Marsh Habitats

Many of these sites are reserves (some National Nature Reserves) where public access is seasonal and limited, therefore details should be sought from the relevant authority before planning a visit.

Broadland Conservation Centre (NNT), Norfolk, TG 359151. Off B1140, NE of Norwich. A trail passes through a range of broadland habitats to the conservation centre overlooking Ranworth Broad. There is an exhibition of the history and natural history of the Broads.

Barton Broad (NNT), Norfolk, TG 3621. NE of Norwich. Access is by boat restricted to main channels across Broad. Reserve consists of Irstead, Turkey and Barton Broads. Broad and fen habitats. Milk parsley established so there is a population of swallow-tail butterflies.

Hickling Broad (see p 69)

Strumpshaw Fen (RSPB), Norfolk, TG 342067. E of Norwich off A146. Strumpshaw and Rockland Broads straddle the tidal reaches of the Yare and this is reflected in their fen vegetation of some saline-tolerant species. Superb bird life and swallow-tail butterfly; observation hides available.

Bure Marshes (NCC and NNT), Norfolk, TG 3216. NE of Norwich off A1151. Extensive fen and wet woodland. Access ashore by permit only, but there is a nature trail by boat at Hoveton Great Broad (TG 322157).

Holme Fen (NCC), Cambridgeshire, TL 205890. N of Huntingdon off B660. Access with permit only. A drained fen reserve which is still managed in the traditional manner. Rare fenland plants.

Wicken Fen (see p 52)

Stodmarsh (NCC), Kent, TR 222607. NE of Canterbury off A28. A rich area of open water and reedbeds resulting from mining subsidence in R Stour valley. Reed swamp and fen communities.

Burham Marsh (KTNC), Kent, TQ 715620. N of Maidstone, off A229. No access off footpaths. River marshes of the Medway.

Redgrave and Lopham Fens (STNC), Suffolk, TM 0479. W of Diss off B113. Rich fen, river and pool habitats. Britain's biggest raft spider, the great raft spider, has been recorded here.

Rye House Marsh (RSPB), Hertfordshire, TL 386100. E of Hoddesdon. Marsh and pools of R Lea. The Lea is a migration route for many passage birds.

Thatcham Reedbeds (Newbury DC), Berkshire, SU 501673. W of Reading off A4. Access by permit. Reedbeds, marsh and wet woodlands of R Kennet. The reserve is part of a more extensive area which makes up one of the largest and most important reedbeds in southern England.

Threave Wildfowl Refuge (NTS), Dumfries and Galloway, NX 7462. SW of Castle Douglas off A75. Marshes along R Dee.

Lochwinnoch (RSPB), Strathclyde, NS 3558. SW of Glasgow, off A760. Marsh and woodland on R Calder. Nature trail by arrangement.

water-lilies or bulrushes, which colonize the margins of water, give way to drier and firmly rooted fen vegetation, and these once wet areas will eventually be colonized by shrubs and trees. The stages in the progression towards alder carr vegetation can be explored via the fascinating nature trail at Ranworth Broad. As vegetation types are directly related to water-levels, a series of distinct zones forms. Reed swamp which fringes the water may include common reed, lesser reed mace and bulrushes, as well as a number of poisonous plants such as cowbane and hemlock water dropwort. Open fen is frequently composed of a mixture of tall, lush-growing plants, including valerian, hemp-agrimony, meadow-sweet, hairy willow-herb, yellow

A pair of nesting mute swans

and purple loosestrife as well as numerous sedges and rushes. Three plants now becoming very rare are the fen orchid, the fen violet and the marsh pea and another, the marsh helleborine, has declined considerably throughout Britain in recent years as a result of fenland drainage. The zone of shrubs and trees will include alder buckthorn, guelder rose and many different types of willow, and in drier parts, alder and birch. Tree and shrub growth is restricted by winter water-levels. Any areas which are flooded during winter remain open fen and cannot be colonized by trees, for trees will literally drown in standing water. Fen vegetation depends on the type of nutrients flowing through the water, and where there is a high calcium content, vegetation will be rich and varied. Where acidic conditions prevail, bog communities dominate, and although

sphagnum moss features largely, other plants such as marsh cinquefoil, bog bean, bog myrtle, bog pimpernel and ivy-leaved bellflower will appear. Insectivorous plants include sundew and butterwort, both of which will trap insects on their leaves. Their leaves will then secrete an enzyme to enable the plant to digest the insect.

Old canals can be rewarding places for the study of wildlife, particularly the unused and less disturbed sections. Most canalized water is nutrient-rich from run-off received from ditches, which includes sewage and other organic matter. The maintenance and cleaning of drainage ditches actually benefit some floating plants, as the removal of fringe plants prevents them from shading out the floaters. There are some species that are now rare in habitats other than drainage ditches and canals, such as the delicate floating water plaintain or the water violet, found in a few areas of the Midlands.

Some ducks of rivers and estuaries:
a the shelduck spends most of its time in estuaries and is able to feed during all stages of the tidal cycle, but it is also found on fresh water inland
b eider and *c* golden-eye are mainly sea-water diving ducks, feeding on shellfish banks in shallow water when the tide turns. The freshwater diving ducks *d* pochard and *e* tufted duck often form composite flocks of large numbers. Both prefer to dive in shallow water, where the tufted duck takes animal food and the pochard is mainly vegetarian

a

b

c

d

e

f

The surface feeders or dabblers form the largest and most familiar group. They include *f* wigeon and mallard, teal, pintail, shoveller and the ornamental and introduced mandarin

Above A part of the very broad estuary of the River Loughor near the Yspitty works. The area is best known for its oyster catchers, which feed in large flocks at the tide's edge. Their vast numbers have come into conflict with the cockle fishers there

Below Milford Haven estuary with an industrial backcloth. Like all major industrial estuaries, this area is heavily polluted – largely from the discharges of the oil refineries and the power station

Where canals are navigable, the wash from the boats and the turbidity caused by their propellers restrict not only the fringe and submerged vegetation from becoming established, but also the numbers of herbivores that would feed on them. If left undisturbed, fish and invertebrates familiar from other river habitats will be found, including leeches, flatworms, great diving beetles, snails and a host of other creatures which are preyed on by perch, roach, tench and the ubiquitous trout, characteristic of sluggish canal waters. Many dragonflies and damselflies prefer very slow moving waters. Water plants would include hornwort, crowfoot, starwort, duckweed and pondweed, including Canadian pondweed, whose swift passage into many British rivers was achieved via the canal network.

Old waterways, such as sections of the Brecon and Abergavenny Canal (the only canal to pass almost entirely through a national park) have aided the survival of a variety of waterway and waterside plants. Apart from the duckweeds and pondweeds, aquatic species will include branched bur-reed and spiked water milfoil. Along the fringe may be found reed sweet-grass, water plaintain, yellow iris and lesser reed mace, while away from the water's edge will be found damp-loving plants such as bistort, marsh woundwort, water mint and water

---

Some Sites to Observe Canal Habitats

(Contact relevant authority for information)

Buckingham Canal (BBONT), SP 728357. E of Buckingham off A421. Disused canal, ideal for the study of wetland plants and insects.

Field End Bridge Nature Trail (CTNC), Cumbria, SD 526850. S of Kendal, unclassified road to Stainton, off A65. Disused section of canal rich in water creatures, amphibians, fish and water and waterside plants.

Cromford Canal (DNT), Derbyshire, SK 333544-350520. S of Matlock, section of the canal between Whatstandwell and Ambergate. The plant life is richly varied and colourful, particularly in the small marshy areas. Many species of insects, birds and fish.

Sapperton Valley (GTNC), Gloucestershire, W of Cirencester off the A419. Parts of the canal are thick with marsh-marigold and yellow water-lily. Waterside woods encourage woodland flora and fauna.

Lapworth Canal Trail (NT), West Midlands, SP 186709-188678. SW of Solihull. Waterside nature trail along Stratford-upon-Avon Canal.

Llangollen Canal (Clwyd CC), SJ 233433. Bankside hedgerows, meadows and marsh areas provide wide range of waterside habitats.

Glamorgan Canal (Glamorgan CC), ST 143803. On the NW outskirts of Cardiff. Canal flows through alder carr and mixed woodland. Varied bankside flora and fauna. Information from Cardiff City Council.

Grand Western Canal (Devon CC), SS 999124. E of Tiverton, near Halberton. Walk along the Canal rich in plant and animal life. One of Devon's few slow flowing waterways.

Daisy Nook Country Park (Manchester BC), SD 921004. On the outskirts of Manchester off A627. Underdeveloped section of this industrial canal, providing refuge for many river flora and fauna.

Etherow Country Park (Manchester BC), SJ 965909. E of Stockport, off A626. Quieter sections of the canal and R Etherow winding through woodland and marsh to provide some sanctuary for wildlife.

Prees Branch Canal (STNC), Salop, SJ 497332. SW of Whitchurch. Unclassified roads off B5476. Disused canal invaded by alder and reed swamps. Fine show of cowslip and marsh-marigold along the banks in spring.

Five Locks Canal (GTNC), Gwent, ST 287968. Disused section of the Brecon and Abergavenny Canal between Cwmbran and Pontypool, bordered by fields and lined with alders. 120 plant species have been recorded from the water and its banks. Monkey flower, an alien invader which has spread through the canal system, is well established here. Breeding site for newts and frogs and toads.

forget-me-not. Stinging nettles grow profusely along canal banks. Although an indication of rich soils, they are a sign of disturbance. If they are abundant on a bank, it is a good indication that they have colonized on the rich spills of dredged material. Once established, they are very difficult to remove or replace with other plants.

The invertebrates and insects which favour this habitat will in turn provide food for waterside birds with which we are already familiar. A pair of mute swans and their cygnets are a common sight on canals. These birds may become tame enough to supplement their diet by begging scraps of food from passing craft, but all too often they will die slowly and painfully after swallowing discarded fishing tackle and lead shot; casualties are high on canals which are regularly frequented by anglers.

The most striking feature of 'natural' lowland rivers is the incredible variety of habitats which accommodate a diversity of interdependent aquatic and terrestrial flora and fauna. Regrettably the impact of man's activities on lowland rivers has not only simplified its ecosystems and changed its microhabitats, but it has sometimes transformed an entire natural river system. Organic and chemical pollution are creating large stretches of sterile waters. All too often moist pastures and meadows that lie in river corridors are drained to make way for agricultural land, affecting the flow and water-table of an entire river system. In heavily populated areas, river systems that escape drainage may become recreation areas creating different problems. Man's impact is now such that the future of the living things with which he *shares* his environment is almost entirely dependent on his tolerance and sympathy. Any hope in reducing the pace of destruction caused by our 'progress' can only lie in the growing public concern for the health and appearance of our surroundings.

CHAPTER 9
Estuaries – the Meeting of Two Water Worlds

Estuaries are the places where the fresh water of the river meets the salt-water of the sea. Carried in suspension in the fresh water are particles of rich organic mud, which were collected from the whole catchment area of the river and carried downstream by the current. They accumulate to form mud banks and flats, which are alternatively covered and uncovered by the ebb and flow of the tides. The continuous flow of rich organic matter creates a habitat abundant in food resources for plant, animal and bird life, but it also presents problems, particularly to invertebrate and plant life, which consequently show extraordinary adaptations to survive in this peculiar environment.

The waters in the estuary area are not homogenous but are layered: fresh water flowing seawards at the surface, heavier salt-water flowing upstream with the tides at the bottom, while between them is a brackish mixture of river and sea-water moving alternately in one direction then another with the rise and fall of the tide. Thus freshwater creatures moving down from the rivers live on the surface of estuarine waters, and these include fish such as bream, roach and three-spined stickleback. Examples of marine animals which penetrate estuaries from the seaward end to feed are fish such as the common goby, grey mullet and flounder, the latter being the only flat fish in Britain which will move some way up through estuaries to live, for a short time, in fresh water. Spawning takes place offshore in spring, and after the eggs have hatched and the larvae have developed into their adult form, the young flounders move into estuaries and then can penetrate far upstream into fresh water if there are no obstructions, such as waterfalls, to negotiate. After two to three years they migrate back to the sea to spawn. Its mottled brownish coloration provides camouflage at the bottom where the flounder settles very still after flipping sand up over its body.

Some molluscs and sea-snails, more usually found along the seashore, may sometimes be found in the mouths of estuaries, and can be observed at low tide. These include the widespread common periwinkle, edible winkle and cockles,

---

Examples of Estuaries Showing Rich Invertebrate Fauna

Morecambe Bay, Lancashire, SD 4070. Sandflats and salt-marsh habitats. Information from Leighton Moss reception centre. Interesting creatures of sandflats include a small shellfish, the Baltic tellin. Enormous numbers of *Corophium volutator*. 160 bird species recorded: a quarter of the country's winter population of bar-tailed godwit, knot, oyster catcher and turnstone come here for food and shelter.

Dee Estuary, Cheshire. *Gayton Sands* (RSPB), SJ 273790. Large area of fresh- and salt-water marshes. *Red Rocks Marsh* (CCT), SJ 204884. Permission needed for access.

Dee Estuary, Clwyd. Mostyn to Ffynnongroew stretch. Access off A548. *Point of Ayr* SJ 1285. Mudflats contain numerous molluscs such as Baltic tellin, peppery furrow shell and sword razor. Variety of salt-marsh plants.

At the mouth of the Looe River in Cornwall. Deep water estuaries are often the sites of fishing harbours, attracting scavenging herring gulls

the microscopic algae with which it is coated. Their shell colour ranges from an attractive bright yellow to olive-green, or brown and purple streaked forms.

The common limpet is another sea-snail species seemingly found on every rock. Even when the tide is out it will continue to graze, very slowly moving about while the rock is wet. It feeds by scraping encrusted vegetation with its horny, toothed 'tongue', or radula, the characteristic feeding organ of all molluscs. The moment there is any danger of it being dislodged or of drying up, its sucker-like grip pulls its shell firmly down against the surface to which it clings, and so it remains until the tide returns.

Acorn barnacles and the common mussel are also fairly versatile and will establish themselves on rocks and boulders where there is a constant supply of microscopic organic debris and planktonic organisms to feed on. When covered by the tide, the tiny beating hairs on their gills create a current to draw in food, which is then trapped on a mucous film on the gills. The waste is sieved through and washed away in the stream of water.

Another interesting marine animal which ventures into estuaries is the commonest of British crabs, the shore crab. It varies widely in colour, usually dark green, but sometimes patterned with paler marks. It is able to tolerate very low salinity levels and is an indiscriminate scavenger, feeding on live prey or rotting carcases, and so it survives well on rock, sand or mud. The creature will often make a tasty meal for a gull, or even the carrion crow, which has learned to hunt in estuaries when the tide is out, but if very lucky, the crab may escape with just a damaged limb. When this happens, a special 'autotomizer' muscle in the upper limb, which is attached to a point in the lower limb just above a preformed 'breaking line', will automatically contract, bending the lower limb so far back that it fractures cleanly along the breaking line so shedding

which seem to be able to live on any type of bottom surface from bare rock to stones, gravel and even on mud. Research has shown that the periwinkle can orientate itself by moving towards and from the sun, and this may be demonstrated in the frequently seen almost oval route tracks in soft mud where the animal has tried to keep to the same position. The tiny, small periwinkle, with its blue-black rounded shell, is also common, usually packed in such numbers in crevices that they cannot be missed. Where strands of bladderwrack seaweed have drifted into the mudflats and become lodged amongst protruding rocks and boulders, the almost globular, brightly coloured round periwinkle may be found on the surface of the weed, where they feed on the damaged portion. A membrane will then seal off the upper limb, and soon afterward the new limb starts to grow.

The more permanent inhabitants of the estuarine environment are the few invertebrate species which can withstand the fluctuating salinity levels of brackish water. As they face less competition from other species for food and living space, having carved out a niche in this unusual environment which only they can tolerate, they can occur in high densities. One such creature is the minute mud-snail *Hydrobia ulvae*. It has a distinct tidal cycle: when the tide is out, the snail feeds on the surface organic detritus and then it burrows into the mud just below the surface. As the tide rises, it resurfaces on a raft of mucus formed from its own glands. The snail becomes attached to the surface film, collecting plankton as it is buoyed upshore by the tide. When the tide ebbs, it gets stranded again in the mud, but this behaviour allows the creature to feed throughout the tidal cycle.

Another common, but less conspicuous,

Though many people assume that the Clyde is lined with heavy industry, long stretches of the estuary lie against attractive open countryside, as between Dumbarton and Helensburgh

estuarine animal is the small amphipod Crustacean *Corophium volutator*. When exposed, it crawls into its U-shaped burrow, only to emerge when the tide is 'in', to forage for edible fragments washed in by the sea.

At low tide a walk across an estuary will usually take one past a multitude of lugworm casts. This creature also inhabits a U-shaped burrow, and from time to time the worm's head will appear up the 'head' shaft to swallow sand mixed with organic matter from the food-rich surface layer. This passes through its gut where the organic matter is digested, the rest being projected through the anus up the 'tail' shaft, creating the coil-like casts on the sand around the burrow opening. The worms are popular bait for sea fishermen who dig them up at low tide. A cone-shaped depression forms in the sand at the spot above the head shaft, telling the bait-digger the direction in which the body of the worm is orientated. This enables him to dig up the worm whole.

Other common animals tolerant of brackish water include ragworms, sludge-worms, some sea-anemones, prawns and shrimps, such as the opossum shrimp, so called because it carries its eggs in a sort of a

pouch until they hatch. This grey inch-long creature forms an important link in estuarine food webs since it scavenges on a variety of organic matter and is itself eaten by many fish and birds.

Spectacular migrations through estuaries are made by fish on their way to feeding and breeding grounds. This is possible because of their ability to control their internal salt and water balance through an enormous salinity range. In bony fish, such as salmon and trout, the concentration of their blood salts is between that of fresh water and sea-water. As the saline concentration of the water is higher than the blood salts of the fish, to compensate the tendency to loose water, they will drink enormous quantities of sea-water and at the same time excrete chloride salts using secretory cells on their gills. In fresh water the reverse happens, and to counteract the large volumes of water taken up the fish will excrete the excess in dilute urine. Salmon, sea lamprey, sturgeon, sea-trout, shad and smelt all swim upriver to spawn, while eels, or 'elvers' swim upriver to feed and mature in fresh water, then return to the sea to breed.

All round the coast of Britain, the immensely rich food resources of the estuarine habitat provide year-round and overwintering feeding sites – and sometimes even breeding sites – for many resident and migrant waders and wildfowl. Many British estuaries are therefore areas of notable scientific and ornithological interest, of which Morecambe Bay, the Wash and the Severn support the largest concentrations of overwintering wildfowl and waders in Britain. Large flocks of other migrant birds may also congregate at estuaries because of the food-rich fresh- and salt-water marshes preceding the flats.

Waders, a term encompassing the large flocks of different species of birds feeding on the tidal flats, include short-legged waders – the ringed plover, sandpiper, dunlin and knot – which feed at the water's edge,

A cluster of molluscs, largely composed of mussels. Large colonies indicate the occurrence of prolific plant plankton, upon which they feed

waiting for the ebbing tide to expose the mudflats, and longer-legged birds – snipe, curlew and redshank – which wade out into the water to feed. The length of a wader's bill will also determine the depth to which it can probe for food in the mud, so that different species exploit different facets of the food supply. Those with shorter bills, for example the ringed plover and knots, can only reach to feed on invertebrates in the surface layers, such as *Hydrobia* snails – an important food source for short-billed waders and also for the omnivorous shelduck and pintail – and ragworms which will emerge to the surface to feed when the tide is ebbing, while longer-billed birds such as the curlew, the godwits and the oyster catcher can reach deeper down into the mud to extract lugworms from their burrows. Some birds, like the oyster catcher, are quick to take advantage of any situation which will add to their food supply. Evidence of beak marks in mud shows how the birds will probe along the lower levels of accumulated mud, and they will even follow a line of human footprints, probing in every one for food usually lying too deep out of their reach, such as peppery furrow shells, as the depth at which this food lies will be somewhat reduced because of the depressions formed by the footprints.

Another factor which reduces competition within high-density flocks is that no two species of wader will feed for the same duration within a complete tidal cycle. During this period of time the curlew will spend about fifty per cent of the time feeding, while the dunlin will spend longer, about seventy-five per cent of the tidal cycle. During the shorter winter days, waders will flock where food is most abundant and feed into the night if low tide occurs during darkness. The oyster catcher will search out cockles at night by walking forward with the tip of its bill near the mud surface until it locates one. Wear and tear on a bill put to such use is compensated by continual growth. By the end of the season, the enormous pre-winter populations of estuarine invertebrates will have been considerably reduced, but since those creatures that will have survived will also have increased in size, the balance is naturally adjusted as the preying waders will consume fewer to fill themselves.

In addition to waders, the estuarine habitat will also support other bird species from large concentrations of overwintering geese and duck, and smaller flocks of whooper and Bewick's swans, to opportunist feeders such as gulls, carrion crow, jackdaw and the occasional heron seeking fish in estuarine pools and creeks. Several species of goose stop at estuaries on their way to breeding destinations in the higher latitudes of the northern hemisphere, and often one kind greatly outnumbers another, depending on the locality. White-fronts from Greenland favour many of the areas frequented by barnacle geese in Ireland, northern Wales and on the coast of Scotland along Dumfries and Galloway, while European white-fronts haunt the Severn estuary and there is a smaller population in north Kent. The east side of Scotland right down to the Wash in England is where largest concentrations of the pink-footed goose flock. The best barnacle goose haunts are those mentioned for the Greenland white-fronts, but the Spitsbergen

---

Some Estuaries of Notable Scientific and Ornithological Interest

The Wash, Lincolnshire and Norfolk. Excellent birdwatching and rich mudflat, salt-marsh habitats. Diversity is added with an area of reclaimed pasture close to the shores of the Wash. Many access points. In Lincolnshire these include *Friskney* TF 5054, *Wrangle* TF 4550, *Boston Point* TF 3939, *Holbeach* TF 4034, *Nene Mouth* TF 4926. *Frampton Marsh* TF 3638, a large salt-marsh, is part of the only unreclaimed area of high salt-marsh showing a fine example of salt-marsh zonation. A wealth of insects, molluscs and other small animals live in the vegetation, providing food for vast populations of birds: the reserve protects the largest colony of black-headed gull. The salt-marsh also draws large flocks of seed-eating birds in winter to feed on the seeds of the vegetation. In Norfolk, access points are at *Terrington* TF 5824, *Ouse Mouth* TF 6023 and *Hunstanton* TF 6740. *Holme Dunes* (NNT), TF 9188, is a large nature reserve covering an area of salt- and freshwater marshes. A nature trail can be arranged.

New Grounds (WT), Slimbridge, Gloucestershire, SO 7205. Lying at the head of the Severn estuary, this area covers nearly 2000 acres of mudflats, salt-marsh and low-lying fields. The Wildfowl Trust has its headquarters here, with the world's largest collection of wildfowl. White-fronted geese are one of the Slimbridge specialities, and Bewick's swan another. Many duck species are also present, mallard and wigeon occurring in their thousands. The estuary by New Grounds is not too good for waders, as the mudbanks are not rich in invertebrates, but in times of passage small numbers do occur, lapwing and dunlin being the largest flocks.

Some Estuaries Showing the Density and Diversity of Estuarine Flora and Fauna

Dovey Estuary, N Gwynedd, SN 6196. Access at Aberdyfi, off A493.

Dovey Estuary, S Dyfed, SN 6094. Access at Ynyslas, off B4353. At *Ynys-Hir* (RSPB), SN 6896, a fine reserve covering woodland, moorland, salt-marsh and estuary habitats. Not only have many waders and wildfowl been recorded in the estuary, but 67 bird species nest in the oak woodland. With this broad range of habitats, the reserve also supports many insects and mammals.

Solway Firth, N Dumfries and Galloway. *Caerlaverock National Nature Reserve* (NCC), NY 0365, and *Eastpark* (NCC and WT), NY 0565, are the principal wintering grounds of the Svalbard population of barnacle geese, as well as numerous wintering waders and ducks. West of R Nith is *Kirkconnell Merse*, NX 9868, the feeding grounds for large

numbers of geese and a flock of whooper swans. Permission must be sought to enter reserve areas.

Solway Firth, Morecambe Bay, S Cumbria, NY 1757. N of Silloth, off B5302 and Port Carlisle to *Rockcliffe*, NY 2763. Unclassified roads off B5307. These areas are more noted for their waders than their wildfowl: oyster catchers, lapwing and knot in excess of 10,000. Golden plover, curlew and dunlin in excess of 5000. During the spring peaks, pink-footed and barnacle geese can exceed 5000.

Ribble Estuary, Lancashire, SD 4026. W of Preston. Footpaths off A59 and A584 on either side of the bank make long interesting walks. The shore has very extensive sandflats and large continuous salt-marshes and therefore there is great diversity and large numbers of waders and wildfowl: pink-footed goose 10,000+; dunlin 25,000+; knot 50,000+.

population overwinters in the Solway Firth. The population of British greylag geese will reside in the Hebrides, but those that are visitors from Iceland will only transit seasonally in Scotland. The brent goose migrates in winter to settle in a string of haunts from the Wash south to the Thames, and of all geese, these most favour salt-water, being more or less confined to salt-marshes where eelgrass, an unusual herb with long filamentous leaves, grows. As a result of the combination of a mysterious disease in the 1930s and the invasion and complete dominance of salt-marshes by a hybrid cord grass, many British eelgrass beds have been destroyed, bringing a drastic decline in the numbers of brent geese, which have been forced to go elsehwere. Where eelgrass beds have become re-established, and as it is illegal to shoot brent geese, the overwintering population is gradually increasing again in those areas.

All other species of goose are also largely vegetarian, except for the barnacle goose

A preening ringed plover. This short-legged wader is distinguished by its black and white head and black-tipped yellow bill

Estuaries in Norfolk, Suffolk and Essex

The combination of several large estuaries, salt-marshes, mudflats and other coastal habitats makes this coastal area one of the most important for waders, wildfowl and other migrant birds. The brent goose favours this area more than any other in Britain because the marshes have large expanses of eelgrass.

North Norfolk Coast. Between Holme in the west and Cley lies an extensive area of salt-marsh, mudflats and estuaries, providing many rich estuarine habitats with not only a steady stream of migrants but also large colonies of breeding birds. Here the dark-bellied brent goose exceeds 5000, while shelduck, mallard, teal and wigeon also appear in their thousands. There are nature reserves at *Cley* and *Salthouse Marshes* (NNT), TF 054451, *Blakeney Point* (NT), TG 001464, *Morston Marshes* (NT), TG 007443, and *Titchwell* (RSPB), TF 750436.

Breydon Water, Norfolk, TG 4907. Great Yarmouth. Extensive mudflats, salt-marsh and low-lying surrounding pasture encourage a small number of wildfowl and waders, dunlin being the largest flocks.

Minsmere (RSPB), Suffolk, TM 4767, and *Walberswick* (NCC), TM 4773, provide the most productive estuarine habitats along the coast of this county because estuaries elsewhere are narrow, or suffer from the disturbance of recreation. Access to Minsmere is by permit only, but Walberswick contains several footpaths, some of which follow old raised river walls, giving useful visibility over the vast reedbeds.

Stour Estuary, Essex-Suffolk, TM 2033. Good access from both shores. Extensive flats of sand and silt and salt-marshes. Eelgrass is abundant in the marshes and *Hydrobia* in the mud, hence there are large flocks of brent geese and shelduck, among many other waders and wildfowl: wigeon exceed 3000, redshank 4000+ and dunlin 10,000+.

Hamford Water, Essex, TM 2225. Lying between Walton-on-the-Nase and Harwich this area is one of mudflats, islands and salt-marshes, some of which have been reclaimed. Wildfowl and waders are present in large numbers during the migration periods. Breeding birds include black-headed gull, common and little terns, ringed plover and oyster catcher. Eelgrass is plentiful, accounting for large flocks of brent geese. Access is best from south side.

Colne Estuary, Essex, TM 0617. Military areas on the west bank prevent access, but one of the best areas for estuarine habitats and fauna is *Colne Point* and *St Osyth Marshes* (ENT), TM 1013. Access to some areas in the reserve by permit only.

Blackwater Estuary, Essex (partly NCC reserve), TL 8606–TM 0010. The N side is the most important part of the estuary as there is a broad continuous belt of mudflats and the shoreline is fringed with a wide area of salt-marsh. Brent goose numbers up to 5000 here.

Crouch Estuary, TQ 9097. There are fewer numbers of waterfowl here than in other Essex estuaries, but as some of the reclaimed area is now rough grazing and salt-marsh there is some representation of waders and wildfowl. Footpaths run along the embankment on both shores. Further west in the estuary lies the *Marsh Farm Country Park* (Essex CC), on the outskirts of Southend-on-Sea, near Hullbridge. Here marshy fields and dykes draining into R Crouch create a diversity of habitats, and winter brings a good variety of birds here, including the east coast speciality, the brent goose.

Leigh (NCC and ENT), Essex, TQ 835850. South Benfleet. Reclaimed marsh, salt-marsh and mudflats in the Thames estuary. The salt-marsh includes at least five species of glasswort and prolific spreads of common and narrow-leaved eelgrass as well as the soft green alga *Enteromorpha*, an important source of winter food for wildfowl. The mudflats are extremely rich in invertebrates and molluscs and this encourages many waders. Rich insect fauna in the salt-marshes attracts many marshland birds as well: uncommon species include the Essex skipper moth and marbled white butterfly.

which also feeds on molluscs. Goose diet consists of a variety of vegetable matter such as young shoots, grasses and reeds from salt-marshes, waterweed, clover and berries and very occasionally insects and other crustacea, probably swallowed simply because these creatures are attached to the vegetable food matter. At dawn or dusk V-formations of geese can sometimes be spotted spilling out of the sky onto potato and stubble fields near estuaries, where they will add the wasted remains of a harvest to their diet.

Temporary duck visitors to the estuary will include the dabblings – shoveller, pintail, teal, wigeon and mallard – which will feed on the richly organic mud or in the murky shallow water, while flocks of diving ducks – eider and golden-eye – will move into estuaries to feed when the tide turns.

More permanent bird residents, mute swans and shelduck, may even nest above high water in the upper reaches and creeks of estuaries where mudbanks have stabilized. Both are able to feed during all stages of the tidal cycle. Shelduck, being essentially carnivorous, will feed either by wading over the mud to pick molluscs and small crabs, or will up-end in water to select shrimps, prawns, larvae, fish spawn and perhaps small quantities of vegetable matter. Swans, though heavy, manage to walk down to the water over mud as their weight is spread over their broad feet. In the water they feed on green algae, but are also partial to stalks, rhizomes and roots of aquatic plants.

One of the most versatile of birds seen in the estuary is the herring gull which, when the tide is 'in', will seek food inland in fishing ports, on farmland and even in refuse dumps. As the tide ebbs, the birds will begin to appear from their inland feeding spots to start hunting on the creatures left exposed and stranded on the estuarine mudflats – crabs, snails, worms, mussels and other molluscs are all swallowed. An oyster catcher 'hammers' open a mussel with its

Lesser white-fronted goose. A smaller edition of its more common cousin, the white-fronted goose

Pink-footed goose demonstrating the goose-step

powerful bill, and although the gull also has a powerful beak it has not yet learned this feeding technique and will instead fly up into the air with its mussel and drop it from a height, usually cracking it open sufficiently to be able to swoop down and chisel out the flesh. Other British gull species in this habitat include common, black-headed and kittiwake, as well as a summer visitor, the lesser black-backed.

Mudbanks build up along the margins of sheltered parts of estuaries where the silt carried down by rivers is deposited and trapped among pieces of seaweed and other debris. On these banks one or two species will colonize to stabilize the mud to form the lower marsh. One of the first plants to colonize is glasswort. It has roots which grow down into the mud and help bind it together, holding it fast against the fluctuating tide. It is also tolerant of the high and variable salinity of the salt-marsh environment. The plant is collected along the east coast of Norfolk, where its salty, fleshy stems are a popular delicacy in salads or pickles. Once this pioneer plant has secured a foothold, other groups of salt-tolerant plants will appear, such as sea plantain, sea purslane, sea arrowgrass, sea milkwort, sea aster and sea lavender. The natural pattern of salt-marshes in Britain is changing due to a recent plant invader: the hybrid cord grass. A new species of cord grass, *Spartina alterniflora*, first recorded in Southampton Water in the 1870s, hybridized with the native English cord grass *S. maritima* to become a most successful early colonizer of bare mud. The hybrid spread fast by natural means, but when its economic value as a mud binder was realized, it became widely planted by coastal protection authorities. However, it has proved to be an aggressive invader and has, as a result, unfortunately destroyed the interesting vegetation of many salt-marsh communities.

Where such destruction has not yet taken hold the above-mentioned pioneer plants will gradually thatch over and stabilize the mud. Each flood tide will bring a new load of silt, depositing it progressively to raise the mudbanks above the water-level, enabling other plants to become established. Thus a zonation can be observed which is

Bewick's swan feeding: one of two species of wild swan that regularly visit Britain in large flocks

related to the duration each zone of vegetation is covered by the sea. Salt-marsh plants flower late in summer when the tides are lower and the marsh is not completely covered, so that wind and insect pollination can occur. Then the pattern of zones can be observed by the colours of their flowers: the pink carpet of seathrift in the upper marsh blends into the lower carpet of the violet-blue common sea lavender which yields further down to the mottled coloration of purple and yellow sea asters.

On a warm day the mudbanks of the estuary will swarm with insects, the most common being tiresome mosquitoes, gnats and midges. Their larvae can be found in stagnant brackish pools. In addition, various terrestrial insects exploit the nectar-sweet

Wyre Estuary Marshes (LTNC), Lancashire, SD 3446. NE of Blackpool off the A588. Ungrazed estuarine marshes especially fine for excellent examples of vegetation zonation. Because of the tidal nature of the Wyre on this stretch, plants and animals show an amazing adaptation to a range of salinity – from living in pure salt-water in the spring tides to dried out brine in the summer months.

flowering plants and grasses growing in the salt-marshes, including wasps, bees, butterflies, moths, beetles and spiders. Butterflies include the grayling, meadow brown and small heath. Their caterpillars feed on a variety of grasses. Sea wormwood and sea plantain are the food plants of the ground lackey moth, found only in north Kent and Essex. The spiders that inhabit this environment, although not truly marine because they do not live in the water, are capable of withstanding submergence at high tide. A typical salt-marsh species is the wolf spider, *Pardosa purbeckensis*, and another, the money spider *Praestigia duffeyi*, which lives on bare mud under the cover of sea purslane, is found locally in south-east England.

Areas of salt-marsh provide excellent grazing areas for cattle and sheep. In East Anglia the sites of estuaries from which the sea has retreated and the salt has been washed away by the rain have been claimed for farming. All Britain's major ports are situated on estuaries and many other industries were similarly sited to take advantage of sea transport. They are also valued sites for oil refineries and power stations. As a result, estuaries are often heavily polluted. Poisonous heavy metals such as lead, cadmium and mercury are discharged into these areas from mining and chemical industries, to be accumulated in some invertebrate, thereby affecting the successive food chain. New threats include atomic power stations which draw in water for cooling purposes and discharge it in huge volumes of hot water, several degrees above ambient estuarine water. In these times of diminishing fuel resources, proposals are being put forward for the construction of estuarine barrages as a means of harnessing tidal power. While any alternative to nuclear power might be welcomed, too many such barrages would transform the estuarine habitat, which is so dependent on the ebb and flow of the tide, while the effects on migratory overwintering birds who have for so long used our estuaries as 'stopping-off' places, would be severe.

An aerial view of the Exe estuary near Dawlish. The estuary is one of the most important wetlands of southern England. The large area of sand and mudflats supports considerable growths of *Zostera* and *Enteromorpha* encouraging many waders and wildfowl, particularly large flocks of wigeon *(Cambridge University Collection)*

CHAPTER 10

Rivers in Danger

The information and opinions contained in this chapter represents a digest of what the authors regard as being the most relevant published material. It does not necessarily reflect the official outlooks and policies of the National Trust.

Threats to the wildlife of our fresh-waterways have existed since human communities first began to disturb the natural ecosystems with their occupancy, farming and fishing operations. But the disturbances were relatively small before the days of industrialization, conurbations and chemical farming. Until the Industrial Revolution Nature was not overwhelmed by the demands man made on the waterways. With the changing farming techniques, the rapid expansion of population and industry in the last century, and the increasing recreational demands of the last few decades many of our freshwater bodies have been quite unable to cope with the speed of degradation. Many present problems are associated with pollution and a river can only deal with as much effluent as its oxygen content allows. Once this point is reached, water becomes deoxygenated and life in the river will decline. There is now growing concern for some of our estuaries, which may be destined to a similar fate.

Pollution and degradation of freshwater sites take many different forms. Until late in the last century most of the lowland rivers of Britain had floodplains in their lower reaches, providing extensive marsh and meadow havens for flora and fauna of all kinds. Comprehensive drainage (sometimes occurring much more than a century ago, as in the East Anglian Fens) has replaced these riverside wetlands with uniform expanses of farmland dissected by ditches. The pace of drainage has shown little sign of decline in recent years and in many places it has been stimulated by EEC subsidy support despite the 25 million tonne grain surplus for which there is no economic market. A survey done by the British Trust for Ornithology in 1982 estimated that in the last few years 18,000 acres of wetlands have been drained annually. The problem has been highlighted in the recent case concerning agriculture and conservation and events involving 5000 acres of the Halvergate Marshes in the Norfolk Broads. For centuries the marshes have been used as grazing land in summer and left fallow in winter to become a favoured habitat for wildlife but it has now become financially attractive to plough up the marshes to grow wheat, a commodity in over-supply which is bought at a guaranteed and subsidized price fixed by the EEC. The traditional British countryside, on the other hand is in very short supply. Although the Ministry of Agriculture and the Department of the Environment rejected a recent application for grant aid to drain the Marshes, the Wildlife and Countryside Act of 1981 provides that where the land has a particular conservation value, ie is scheduled as a Site of Special Scientific Interest, occupiers and others can receive compensatory payments. The cost of protecting the countryside in this manner could become prohibitive, particularly if the price of compensation is to be based on the value of the land if under cereal with all the subsidies which would thereby accrue.

Extensive areas of the Halvergate

Geometric field drains – like this one in the Fens – interfere with the complex natural system of lowering the water-table, reducing marginal habitats, and can spread farm chemicals through the river system

Marshes have been lost, as described in Chapter 3. Similar threats exist in other key lowland British marshland sites, namely the North Kent marshes, our largest surviving wetlands, the Somerset Levels, the Ouse/Nene Washes in the Cambridgeshire Fens and the 2000 acre Lower Derwent Ings in East Yorkshire. In Ireland, Rahasane Turlough in Co Galway is also under threat, and the commercial exploitation of peat is also affecting the natural balance there.

While some record has been made of the loss of larger and obvious species of plant and animal life through such destruction, a myriad of smaller life forms also suffer. Wintering waterfowl which once would have gathered in great numbers where there are now ploughed fields are absent or unusual as nesting birds over most of inland

lowland Britain. Fenland breeding birds, such as the bittern, the marsh harrier and the bearded tit, which would once have been quite abundant, now confine themselves to a few sanctuaries which are protected nature reserves. In 1978 P. D. Round compared the population of breeding birds in two areas on the Somerset Levels which were virtually identical except that West Sedgemoor, east of Taunton, was 'natural' rough meadows used as summer pasture, whereas the other, Witcombe Bottom, near Langport, was pump-drained in 1977. The results are shown overleaf.

There were once forty-one species of dragonfly in Britain. Four species have become extinct since 1953 and three of these are considered to have been lost as a result of the lowering of water-tables. This lowering of water-levels has caused permanent pools to dry out in spring and summer, when the nymphs emerge to become adults, breed and

Species	West Sedgemoor No of species	Witcombe Bottom No of species
Lapwing	13	5
Snipe	16	0
Curlew	1	0
Redshank	4	0
Skylark	28	15
Grasshopper warbler	2	0
Sedge warbler	11	0
Meadow pipit	16	0
Yellow wagtail	1	4
Reed bunting	15	1

(P. D. Round, 'An Ornithological Study of the Somerset Levels 1976-77', unpublished report by the Royal Society for the Protection of Birds and the Wessex Water Authority, 1978.)

The smooth newt needs water for breeding, and like other amphibians has lost many of its breeding sites

Stickleback. The ten-spined stickleback is another creature threatened by the disappearance of ponds and ditches

lay their eggs. The Norfolk coenagrion has almost certainly become extinct through the desiccation of its habitats. Common river species such as the banded agrion and the demoiselle agrion, while still widespread and able to withstand minor river engineering, are also declining as mechanical clearance operations reduce pools and ditches which form in the shallower margins of a river-bed.

The reduction in fen habitats has led to the loss of many bog, fen and marsh plants found along the river corridor. A plant which was once found in many such localities scattered throughout Britain was the marsh helleborine. Its decline and much reduced distribution is a direct result of the destruction of its natural habitat. The various forms and hybrids of marsh orchid and early marsh orchid have also suffered in this way. More common plants such as the marsh-marigold, once found in virtually any moist corner, is now also far more restricted in its distribution, being confined to niches on river margins and boulders.

The advent of piped water supplies for

livestock has also reduced the need for riverside ponds and ditches, and the lack of such traditional damp sites during the spring breeding season of amphibians is taking a serious toll on frogs, toads and newts. The ditches and pools that do still remain may be contaminated with rubbish, such as the ubiquitous polythene fertilizer sack or empty drink cans, farm effluents, pesticides and aquatic herbicides. Effluents are a source

Mute swan. Swans are dying painful and lingering deaths at the rate of over 3000 a year from lead poisoning

The Grand Union Canal near Marsworth, in Buckinghamshire. The canal, originally built to supply industry in the Midlands, is now an avenue for motor-powered boats. As in many other areas under heavy recreational pressure, the wash from boats and turbidity caused by the propellers restrict the development of water and fringe wildlife

Here, the Ouse is an attractive feature of York, but downstream it is seriously polluted (see page 197).

a The Thames at Mill End. Although the river is under pressure from industry, domestic use and recreation, pollution is carefully controlled and some quieter sections of the river support much wildlife

b Wicken Fen Nature Reserve. Early drainage of farmland was once carried out by wind pumps, but now much time and effort are spent in pumping water back into the Fens in order to maintain water-levels to protect the Fenland species

c Reed cutting on the nature reserve at Wicken Fen produces a crop for the National Trust while at the same time maintaining an open reedbed, encouraging the colonization of other plants and helping water flow

of organic enrichment and in still waters they lead to the danger of massive algal blooms which will have the effect of smothering higher plants and eventually cause the death of fish, such as stickleback, and other water creatures through deoxygenation of the water, while pesticides and aquatic herbicides deform or kill. It is known that even very low concentrations of pesticides cause deformities in frog tadpoles and so hinder their development into the adult stages. Research published by the Nature Conservancy Council allows us to compare the biological value of traditional water bodies and their modern equivalents.

As we have mentioned in previous chapters, a serious side-effect of recreational activities concerns the painful death of birds which results from lost or discarded fishing line,

	Permanent ponds and ditches	Temporary ditches and piped water
Mammals	2	0
Amphibians	5	2
Fish	9	0
Dragonflies	11	0
Snails	25	3

('The Conservation of Farm Ponds and Ditches', Nature Conservancy Council, 1982.)

hooks and lead shot. Swans, our 'royal' birds, are dying at the rate of about 3000 each year as a result of eating such shot. Discarded weights are swallowed, usually accidentally, together with the grit that the

Perhaps a third of these Bewick's and whooper swans will die from lead shot

creature needs in its gizzard to grind up its food. In a recent article in *Natural World*, Gareth Huw Davies claims that a lethal dose can be as few as eight pellets, which, when ingested, work as a poison which takes effect slowly over a few agonizing weeks, weakening the bird's muscular contractions so that food is restricted from passing into its stomach. Consequently the swan will starve to death, often with a neck full of food. Dr Mike Birkhead of the Edward Grey Institute of Field Ornithology at Oxford University estimated that in England and Wales discarded weights were found at an average distance of one for each 6 inches of river-bank – amounting to about 250 tonnes each year.

Fishing tackle is another hazard. In 1970 members of the Young Ornithologists' Club collected an average of 900 feet of discarded line per mile of river-bank and found 42 birds of 15 species killed by it.

Evidence of swan deaths on the Thames has been provided by Captain John Turk, Keeper of the Queen's Swans for twenty years. As reported in the *Guardian* (18.7.85), in 1984 only 32 cygnets were found between Windsor and Pangbourne, 20 fewer than in the previous year, while 30 years ago the same stretch of the river yielded over 1000 cygnets. In July of 1985, following mounting public concern over swan deaths, the junior environment minister, Mr William Waldegrave, announced the phasing-out of lead weights, a measure which he hoped would lead to the restoration of the national stock of swans to a total of about 19,000. In order to allow anglers to adopt substitute weights, the ban on lead weights will not become effective until the start of 1987, though it is immediately effective upon applicants for fishing permits in the Royal Parks and model by-laws have been prepared for water authorities and others who seek to follow this example.

Pleasure-boating enthusiasts can also create major problems, especially where attempts are being made to establish avenues of navigation through quiet waters where wildlife systems are virtually intact. The Derwent is under such threat because the Derwent Trust for Navigation is seeking to re-establish navigation, even though the river is so important that the Nature Conservancy Council is in the process of scheduling the whole river, beds, banks and all, as a Site of Special Scientific Interest. Although navigation once existed on parts of the river over fifty years ago, it was extinguished by Parliament in 1935 and since then the river's wildlife has had the chance of existing free from many of the pressures experienced on other British rivers. The Derwent Trust for Navigation has not accepted the 1935 decision and has gone as far as to file evidence against four riparian owners in the High Court, basing their case on the fact that they are acting in the public's interest by seeking to prove navigation still exists on the river. Attempts are being made to raise money to fight the case against pleasure-boating, but if this fails then there will be no stopping the Inland Waterways Association whose plan it is to open up all potentially navigable waterways.

A major source of threat to our rivers is the pollution from millions of gallons of sewage-laden water that flows into the rivers daily. The River Tyne, for example, receives 70 million gallons of untreated sewage every day, which is about the same as the volume of freshwater flow down the river during its drier minimum flow periods. This is because only five per cent of the sewage from the one million Tyneside inhabitants is treated before it goes into the river. This is added to waters often already rich in dissolved minerals, such as nitrates and phosphates from fertilizers applied to the land; slurry from cowhouses and pig farms, and other effluents from our food and agricultural industries. Nitrate pollution in drinking water is dangerous as there is

The Aire is one of our polluted and lifeless rivers

processing plants, fish-meal factories and breweries are all released into our rivers, causing further eutrophication. Silage liquor is a particularly harmful pollutant as a 400 tonne load of unwilted silage compares with a day's untreated sewage from a town of 150,000 people (Nature Conservancy Council). Yet such agricultural dumping may not be properly monitored. In slower stretches of rivers and ditches such enrichment will inevitably lead to massive algal blooms and deoxygenation. Domestic sewage increases the saline content of the water, and a side-effect of this is that salt-tolerant plants, such as fennel-leaved pondweed, become widespread at the expense of other plants, such as water crowfoots.

An unlimited supply of organic waste matter will also stimulate the proliferation of certain riverine fauna at the expense of others, resulting in an impoverished representation of the species characteristic of the former, more healthy habitat. Changing food material will affect the competitive status of some species, such as detritus feeders, scavengers and bacteria that break down a wide range of organic material, leading to their population explosions. More oxygen will be used for respiration by their large numbers, thereby progressively lowering its levels, so that organisms with high oxygen requirements, such as stone-fly and mayfly larvae, will eventually disappear. The changing fauna able to exist in lower oxygen levels will be those particularly adapted to obtain the element, such as the rat-tailed maggot, so called because of its telescopic breathing tubes, and the bloodworm, which contains oxygen-carrying haemoglobin. This in turn will determine the population of predators, favouring only those which can exploit these food resources. Fish populations are considerably reduced if their nursery and

evidence of its association with stomach cancer and blue baby disease. But the already high levels in British rivers will rise as farmers continue to use fertilizers. Journalists Sarah Helm and Bryan Silcock in their recent report (1984) claim that the government is stalling on the nitrate problem for fear of delivering another blow to farmers after the recent milk quota ruling, and in fact the Department of Agriculture is actually advising them to increase the use of nitrate fertilizers to save feeding costs. As a result, water authorities have to work on limiting the nitrate level by introducing removal schemes. To be able to implement the new EEC anti-nitrate pollution rules the Anglian authority alone estimates the cost at £32 million. In the last thirty years, our waterways have not only had to contend with ever increasing organic pollution, but also the dangerous and poisonous pollutants of oil refineries and other fuel plants, chemical works and paper mills.

Effluents from slaughter-houses, milk-

breeding grounds lie in such deoxygenated waters. Not only is there a lack of food and oxygen for adult fish, but also for the development of their larvae. The profusion of bacteria in these waters is also injurious to the hatching of eggs and the survival of larvae.

Organically polluted sluggish rivers flowing into estuaries stimulate the overproduction of phytoplankton, which may become so thick that the filter-feeding mechanisms of estuarine creatures may become clogged. Also, the waste products of such blooms may be poisonous to other marine life. Dinoflagellates, single-celled, phytoplanktonic organisms, produce a particularly virulent type of nerve toxin most hazardous to larval and adult fish, and birds.

The chemical pollution of our inland waterways from industry is perhaps an even greater menace because of the indestructability of the pollutants, and industry is tending increasingly to expel its waste products, via our estuaries, directly into the sea. Polychlorinated biphenyls (PCBs) – used in the manufacture of paints and varnishes, as softeners in plastics and for improving electrical insulation in cables – like pesticides, are extremely stable and persist for long periods of time without biodegrading into a harmless form. As quantities have built up over time in water and sediments some sensitive plants and animals have been killed outright. However, some invertebrates have adapted the ability to concentrate them many thousands of times and store them in their fatty tissues (and it is now known that several estuarine molluscs isolate heavy metals by enclosing them in vesicle appendages). As a result other species higher up in the food chain have been poisoned. Some of the 12,000 guillemots which died in the Irish Sea in the autumn of 1969 were found to contain high levels of PCBs. Laboratory experiments have shown that these chemicals can also

Carelessly discarded rubbish and other pollutants in the Aire, as well as being unsightly and unhealthy, are hazardous to wildlife

Dead pike, a victim of pollution

affect the sex hormones of animals so as to render them infertile, and so it may not be entirely coincidental that the fertility rate of guillemots at breeding sites on the British west coast is dropping. It is an awesome thought that even if such chemicals were voluntarily withdrawn from usage today, because they break down so slowly, it will be many years before traces of them finally disappear from the environment.

Other poisons released into waterways include sulphite liquors from pulpmills, phenols and cyanides from chemical manufacture, acid and iron from the production of titanium dioxide, and other heavy metals such as vanadium, manganese, chromium and lead. Copper and zinc from galvanizing discharged into estuaries may prevent Atlantic salmon reaching their spawning grounds, for it is now known that they will avoid waters containing these pollutants even at very low concentrations. Discharges of mercury compounds, used as fungicidal seed dressings, in paper-making and as catalysts in the manufacture of PVC are hazardous not only to freshwater and marine life, but have also been responsible for human death. At Minimata in Japan, between 1953 and 1961, forty-three people died of mercury poisoning and sixty-eight

were disabled. In that outbreak the level of mercury contamination averaged 8.0 parts per million. It is not surprising, therefore, that there is now growing concern being voiced by Scottish environmentalists about the menace of mercury in the River Forth, where the prevailing maximum has been recorded at 6.6 parts per million. Recent research by the Forth River Purification Board has shown that the level of mercury contamination of fish and shellfish taken from the water near Grangemouth is up to twenty-two times above the 0.3 parts per million acceptable by the EEC. The inorganic mercury is converted into highly toxic methylmercury by bacteria which live in the gut of fish. The danger of mercury and other heavy metals is now being recognized, but nevertheless discharge

The Ouse is now becoming seriously polluted between Selby and the Humber estuary

continues in industrial effluent. Such long term accumulation and concentration by lower water organisms will either kill the lower life forms or destroy those higher up in the food chain.

Mining also pollutes rivers and lakes. The unwanted materials from china clay digging and coal-mining smokeless fuel plants have found their way into rivers. In 1969 approximately 1.4 million tonnes of micaceous residue from china clay workings were produced annually in the St Austell area of Cornwall, of which fifty per cent was discharged into the St Austell River and twenty-five per cent into the Luxulyan

Opposite An aerial view of the Tyne at Newcastle. Another one of our dirtiest river stretches *(Cambridge University Collection)*

An aerial view of the Thames in London. The river used to be one of Europe's filthiest industrial rivers, but in contrast to the Tyne, it has now been transformed into a river inhabited by many fish and birds *(Cambridge University Collection)*

River. Where the residue from the waste was suspended in water it robbed the water plants of light and restricted their growth, which in turn affected the insect larvae that depended on them. The material when it settled also smothered the spawning grounds of fish. Although the local river authority received payments from the companies concerned towards maintenance of these rivers in return for their continued use, mounting public concern and protest urged the mining companies to look for ways of eliminating disposal into the rivers. A scheme begun in 1975 disposed of the

Water Under the Whitehall Bridge

Fred Lester, the deposed director of scientific services at the Severn Trent Water Authority, once described the Control of Pollution Act as a 'masterpiece in the art of flexibility'. He was not being complimentary. Here is the timetable of promises for the sections dealing with water pollution:

July 1974: Royal assent for the act.

December 1974: Ministers reveal that they will not introduce the act all at once.

February 1975: Gordon Oakes, a junior environment minister promises that the water sections will be phased in from autumn 1975 to summer 1976.

August 1975: A change of plan. Denis Howell, minister of state for the environment, announces another postponement – this time indefinite. He blames government spending cuts.

April 1978: Howell announces a new timetable. There will be a two-phase implementation, to be completed by the end of 1979.

May 1979: Conservatives win the general election. New ministers at the Department of the Environment, Michael Heseltine and Tom King, announce a new examination of the water provisions of the act. 'There is a feeling at ministerial level that it will cost a lot of money,' say civil servants.

February 1982: After almost three years of review, King announces a five-part programme with the 'maximum use of transitional provisions'. The first parts will be 'brought into force in July 1983'. The rest will be in force by 1986. But King reveals that, eight years after the act was passed, there are to be new consultations with industry about how the sections should be implemented.

July 1983: No sign of promised implementation of the first parts of the legislation.

August 1983: Civil servants explain that there has been another 'slippage' in the programme. But they reveal more details about the 'transitional provisions'. Most discharges that escape existing controls will be exempt from the new ones. The remainder, discharges to the Mersey estuary and of certain toxins, such as mercury and cadmium, that are blacklisted by the EEC, will be given 'deemed consents.' They will be allowed to carry on polluting at existing levels. The civil servants say that 'the exemption order will not be varied by the Secretary of State until public consultations have been undertaken on the next round of priorities'.

The new date for the presentation of the ministerial regulations, which will give companies up to a year to apply for their deemed consents, is November 1983.

November 1983: Civil servants reveal that 'unfortunately our timetable has been slipping yet again because of the extent and the complexity of the subordinate legislation.' Now February will reveal all.

March 1984: As *New Scientist* goes to press, the civil servants and their new Secretary of State, Patrick Jenkin, are still silent.

New Scientist, March 1984

waste by backfilling into disused pits and special disposal reserves. This method of disposal could be implemented without largely sterilizing the reserves or damaging vast tracts of landscape because advancing technology enabled treatment of the waste, and micromineral separation techniques enabled the operators to recover more clay from the clay matrix and thus reduce the total amount of waste to be disposed.

Smokeless fuel plants are notorious polluters. They heat the coal to very high temperatures to drive off the gases and tars which would otherwise be emitted during the domestic use of the coal. The substances emitted include ammonia, phenols and cyanides and are extremely expensive to contain. As well as polluting the air, they find their way into rivers as effluent. In the attempt to treat the effluent and reduce pollution, the substances are being poured over coal spoil heaps, which absorb some of the pollution. However, when the tips become saturated, the solution will

Britain's Dirtiest Rivers

Rother	Chesterfield to Rotherham
Darwen	Blackburn
Roch	Rochdale to Bury
Sankey Brook	St Helen to Warrington
Alt	Aintree, Liverpool to the sea
Ouse	Selby to Humber estuary
Tame	Wednesbury to Birmingham
Yare	Norwich
Ebbw	Newport
Sirhowy	Blackwood
Lostock	Croston
Tame	Hyde to Stockport
Skerne	Newton Aycliffe to Darlington
Taff	Merthyr Tydfil
Dane	Middlewich
Irwell	Bury to Manchester
Glazebrook	Leigh to Cadishead
Aire	Bradford and Castleford to Howden
Dearne	Barnsley to Mexborough
Red	Camborne to sea
Mersey	Salford to Liverpool
Douglas	Mawdesley
Don	Stocksbridge to Doncaster
Orwell	Ipswich
Tees	Eaglescliffe to Tees Bay
Tyne	Newcastle-upon-Tyne to Tynemouth
Manchester Ship Canal	Manchester to Ellesmere Port
Lwyd	Blaenavon to Cwmbran

Fred Pearce, 'The great drain robbery'. This first appeared in *New Scientist* (1984), London, the weekly review of science and technology.

Preening, as demonstrated here by a pintail, is a vital activity of waterfowl, but when their plumage is contaminated by oil, fatal internal damage may follow

eventually find its way into rivers. The Doe Lea, below Bolsover in Derbyshire, is affected in this way and is, consequently, one of the worst polluted rivers in Britain. In a survey carried out in 1980 in England and Wales, rivers were awarded points for the amount of animal and plant life they contained. The Doe Lea managed three points in comparison to the 239 points scored by the River Test in Hampshire!

Although the 1974 Control of Pollution Act and the older stringent by-laws are in existence to curb the discharge of pollutants by riverside industry, Fred Pearce (*New Scientist*, March 1984) considers that big, influential companies have adversely influenced legislation to clean up Britain's dirtiest rivers and estuaries. In England and Wales pollution control is mainly the responsibility of river water authorities and in Scotland of river purification boards and county councils (see Chapter 11). The 1974 Act was intended to bolster the anti-pollution powers of these authorities. In addition, for the first time the laws were to be extended to estuaries, tidal rivers and coasts. In the last thirty years, many of Britain's worst industrial polluters have moved to her coasts and estuaries where, as well as exploiting the wide open spaces and deep berths for big ships, they have enjoyed an absence of pollution control. Due to the many obstacles in implementing the 1974 Act, today many coastal areas are still exempted from the laws of the Act, or industrialists are given 'deemed consents' to carry on disposing of their wastes as before. Ten years on, 2560 miles of Britain's rivers too are as dirty as ever and some continue to get worse because anti-pollution measures are absent or not enforced.

Finance is the biggest obstacle. Extra funds are needed to keep pace with increasing river pollution. In practice, spending cuts have often caused reductions in the investment by authorities on existing sewage treatment and anti-pollution measures rather than allowing any improvements. For instance, the Anglian Water Authority admits the need for research to analyse metals in saline waters, but even the fairly basic information such as: 'the effects that polluting discharges are having on the estuaries, the adequacy of the treatment that the effluents discharged to these waters have received, or the extent to which they need to be treated', is not yet at hand. Water authorities, such as that of the River Tees, have simply had to drop their commitment to 'the restoration of the River Tees to its former unpolluted state . . . because in the current economic climate . . . industrialists may be discouraged from coming to Teesside due to the effluent standards being too high'. In addition, if existing industries flout the law by disobeying limits set for their discharges, water authorities, such as that of the Tees, have no legal backing to force industrialists to reduce and control their discharges. Until 1983 local councils used to appoint the members to the water authority boards. Now these members are made up wholly of ministerial nominees. 'And ministers have shown a penchant for putting men from the top British companies onto the authorities . . . [thus] . . . they may have an effective veto on any proposals put forward by the authority's scientists'. Although there was much enthusiasm within Whitehall to clean up Britain, Fred Pearce claims that this zeal evaporated as soon as the bill was passed 'and since most of the reforms in the new Act required ministerial decision to activate them, Whitehall's coolness has been fatal. The timetable of postponement is long and tortuous.'

Among Britain's many estuaries, the Mersey and Tees are the most polluted. In the 1930s the former offered a living to many fishermen, but today it is too dirty for fish to survive. The Humber estuary, although less polluted at present, is a growth area and a strong indication that there is an urgent need to curb pollution in the area is demonstrated by the fact that although fish survive in the estuary, many species have disappeared, particularly between Grimsby and Immingham. Here a report for the Greenpeace Environmental Group described how pollution from the production of titanium dioxide, widely used as white pigment in the paint industry, has caused serious acidification, with pH levels as low as 2.5 near the discharge pipe – pH is the measure of concentration of hydrogen ions in a solution; an acidic solution has a pH less than seven, while an alkaline solution has a pH greater than seven, and pH seven is a neutral solution. Shore crabs cannot tolerate acidity levels below pH six and barnacles similarly stop feeding. Acidic solution will also seriously impede the development of herring eggs (*New Scientist*, December 1983). More recently, several large estuaries

have become sites of power stations for electricity, and oil refineries, ... while the Fal estuary in Cornwall sees another example of an environmentalist's dilemma. A local container firm is attempting to raise money to convert the estuary into a roll-on roll-off container port. This would involve filling in more than half the estuary. If this happens more than 1000 new jobs might be created in an area of high unemployment. However, those who see it as potentially a major ecological disaster fear severe oil pollution, widespread flooding as a result of impeding the tidal flow in the estuary, and the loss of many life forms in the disturbed river-bed and estuarine environments, leading to the severe destruction of the ecosystem of interdependent creatures. Dr Roger Burrows of Exeter University, as quoted in *The Observer* (1 September 1985), believes that 'The seaweed, the crustaceans, the fish and the birds are all part of a complex eco-chain which would be shattered if the project goes ahead.'

Power stations draw in water for cooling purposes and return it as warm water, several degrees higher in temperature. This can have several harmful effects. Life in the vicinity of the outflow may be killed due to thermal shock on encountering the much warmer waters. Where temperature has been raised, the oxygen carrying capacity of water is reduced, and this hastens the decay of organic matter; heated water discharged in an area already polluted can have disastrous effects.

The continuous discharge of low-level radioactive wastes from coastal nuclear waste processing plants, such as Sellafield, formerly Windscale, in Cumbria, may also be causing damage to life in the vicinity of the discharge point. It has been shown experimentally that low levels of radiation can cause chromosomal damage with subsequent growth abnormalities in both invertebrates and vertebrates. There is some evidence to suggest that exposure of fish to

Power stations are widely regarded as the main contributors of acid rain

radiation can reduce their ability to tolerate changes in temperature and salinity. Some algae and shellfish have the ability to concentrate radioactive elements in their tissues, so that their radiation levels may be many thousands of times higher than the surrounding waters. Existing regulations which have set acceptable dose levels of radiation have attempted to take account of the radioactive accumulating ability of many organisms. But the ecologist J. David George considers that it is yet too early to set acceptable dose levels: firstly the numbers of species so far examined are too few in relation to the actual numbers occurring in coastal waters, and secondly the subtle nature of radiation damage is such that the dose levels should be based on the statistical analyses of longer-term experiments.

Oil seeps naturally from the sea bed in some areas of the world, but the vast majority of oil present in the oceans today has been introduced by man, either deliberately, when discharged by vessels illegally flushing out their oil tanks at sea, or accidentally, when major marine disasters

have taken place: in 1981, 541 spills were reported around the British coast, about half involving over 100 gallons. In addition, untreated sewage contains tars, fuels and lubricants. Where they are not deliberately discharged into estuaries, oil residues can be found floating into these areas as a result of tidal action. Consequently, estuarine life will be affected either chemically or mechanically.

Thousands of sea birds are killed every year as a result of oil pollution. The worst sufferers are those that spend most of their life on or on and just under the surface of the sea and are unable to escape the effects of oil on their plumage, which prevents them from flying and eventually leads to their death by drowning or by starvation. In their efforts to preen themselves, an essential and time-consuming activity of all water birds, they ingest vast quantities of oil, thus damaging their digestive systems. Ingested oil is known to cause degenerative changes in the liver and kidneys. There is also evidence that birds affected by oil have low reproductive potential. When tens of thousands of sea birds have been killed by just one disaster, such as that of the Torrey Canyon in 1967, and some vulnerable species of diving duck, such as scoters and guillemots, lose vast numbers every year, there may well be a danger of the extinction of such species as a result of oil pollution. Other marine animals also suffer when, for instance, the filter feeding mechanisms such as those of some molluscs become so clogged with oil that animals die slowly of starvation, or when air-breathing mammals such as seals and whales find their air passages and lungs have become obstructed, preventing gaseous exchange.

Although some components of oil are gradually biodegraded, the most effective detoxifying mechanism is dispersal. Dispersal of oil into small droplets permits microbial action to proceed more rapidly. Natural biodegradation and dispersal, however, are immensely slow. Hence man speeds up this process by using various types of detergents – but they too can create hazards. The emulsification of oil when dispersed only distributes it more evenly through the environment, so that organisms quite some distance from the source of the spillage will also be affected by the substance. If, however, the wrong detergent is used, such as that poured directly onto beaches in south-west Britain in the wake of the Torrey Canyon disaster, it can result in the death of marine and estuarine life. Long after the disaster it was common to see rotting limpets and other molluscs where the detergents had been used. They also seemed to cause crabs to lose their limbs. Behavioural patterns of marine animals were also modified as a consequence. The dispersant-laden sea-water inhibited the normal climbing responses in winkles and topshells. Other animals which usually remained buried in the sand at low tide, such as *Hydrobia ulvae*, came to the surface, making them very susceptible to predators.

Acid rain is now featuring as a major pollutant of Britain's rivers and lakes. A recent report from the House of Commons Environment Committee identifies acid rain as a crucial contributor in damaging fish stocks, which have fallen due to excessive acidity.

When fossil fuels are burned by factories and vehicles, sulphur dioxide is released along with other pollutants. The term 'acid rain' describes what happens when these chemicals are taken up into the clouds, to fall as dilute sulphuric and nitric acid in rain, often in areas far from their source. Britain is accused of being the main European exporter of such pollution to other countries; thousands of lakes in Scandinavia have been poisoned as a result. Britain's rain is commonly 100 to 150 times more acid than natural rain, and in Bush in Scotland it has been calculated as 600 times more acid!

Fish populations in British rivers and lakes have been drastically affected. Particularly dangerous are snow-melts, when chemicals deposited in frozen water are suddenly released in large concentrations in the spring thaw. The resulting acid levels are thought to be lethal to some newly hatched fish. In Britain a serious decline in fish population has been reported in Scotland, Wales and the Lake District, where there are 'unexplained' fish deaths coinciding with snow-melts.

In September 1983 the *Observer* reported on the findings of unpublished reports of Norwegian scientists who surveyed the streams and lakes in Scotland for pollution. The report revealed that almost all lochs and streams in Scotland are being turned acid. The worst affected areas are those with shallow soils and granite rocks, which cannot neutralize acid. Thirty of the forty streams sampled in the Galloway area were acid enough to kill water life and seven of the rest are in danger.

Buttermere, Cumbria. One of several lakes at risk from acid rain. The North West Water Authority is included amongst organisations and researchers involved in studying the problem.

The North West Water Authority is investigating the chemistry and biology of approximately 100 rivers and lakeside points throughout the Lake District. The results so far indicate fish mortalities in the Rivers Esk and Duddon, usually after heavy periods of rain. Buttermere and the streams leading into it are also in danger.

Most of the streams, rivers and lakes in Wales are also affected by this pollution, some most seriously, such as the important mountain waters of Dyfed, where most fish have been killed as a result. In fact, the Welsh Water Authorities are seriously worried that the whole of upland Wales from Snowdonia to Carmarthen could be affected.

Rivers and streams on peaty moorland surrounded by coniferous forests are much more vulnerable than identical waterways in open moorland because tall pine trees tend to intercept airborne acid gases and particles.

These acid deposits are washed off into the soil to react with aluminium and other metals and are carried away into local streams and rivers when it rains. Catherine Caufield of the *Sunday Times* recently reported the case of a farmer receiving an £85,000 subsidy to turn a 900 acre sheep moor into a conifer plantation despite protestations from the Welsh Water Authority that the plantation would lead to acid rain poisoning the tributaries of the Irfon River, killing the brown trout and affecting local drinking water. The desire for such conifer plantations, although a poor investment – the average loss of private forestry in England and Wales is at about £9.32 per acre – may perhaps be explained by a range of attractive if intricate tax concessions in this form of land use.

The Commons committee recommends that Britain should immediately join those countries such as West Germany, France and Scandinavia which have committed themselves to a thirty per cent reduction in the emission of this gas between 1980 and 1993. The EEC would prefer a sixty per cent reduction. However, both the government's Department of the Environment and the Central Electricity Generating Board (the latter having been singled out for blame as the main producer of sulphur dioxide, at fifty-five per cent, in some reports) suggest more research is needed before decisions can be taken. The CEGB also claims that costs of cutting sulphur emissions would increase the cost of producing electricity from power stations by as much as ten to fifteen per cent, and this cost would have to be passed on to the consumer. In a letter to the House of Commons Select Committee, dated June 1984, the Assistant Chief Agent of the National Trust wrote: 'With regard to environmental effects the Trust has not detected any general environmental damage which could be attributed to acid rain on its farmland, woods or open spaces properties … The Trust's Adviser on Conservation and Woodlands while aware of the current reports of massive damage to forests on the continent believes the evidence so far collected points to a complex combination of atmospheric factors which are less likely to occur in the United Kingdom. Nevertheless the decision to review the available evidence is welcomed and the Trust would be pleased to make suitable properties available for that purpose if it would be of assistance'. Britain is unlikely ever to join the other nations pledged to reduce sulphur dioxide emissions, but it is hoped that new technology and the changing nature of the industry will reduce emissions to some extent.

In his effort to improve his own lot, man has sometimes deliberately, and at other times accidentally, unbalanced systems of life which have taken millennia to evolve, sometimes destroying ecosystems to the extent that they can never return to the pre-interference state. Although public awareness has been awakened and the concern for our environment is greater than ever, these times of hardship and competition for new industries and jobs encourage the shorter view. Past and present experience suggests that little improvement can be expected in the quality of our highly polluted rivers, although the success story of the Thames shows what can be done (see Chapter 11). 'High tech' factories claim that a clean environment is an incentive for their location, but such firms are not always job-intensive. Some progress might be made in reducing industrial pollution if more companies were encouraged to use some of their profits to reclaim recyclable materials from the effluent, although no doubt many would claim this is not economically justifiable. As a result many of our estuaries, rivers and streams will continue to receive the waste products of industry, thus becoming stinking and virtually lifeless sites. Questions of pollution control cannot always be postponed until a more convenient time. Fish and plant life will not return as soon as contamination ceases. Some have gone for ever, many are poised to follow. One of the saddest dimensions to the problem concerns the fact that, while it is often asserted that members of the public will not be prepared to pay a little more in order to produce much lighter levels of pollution, the man and woman in the street are never given the choice. The authors suspect that were more people aware of the great issue at stake, human decency and respect and concern for the natural world would favour the necessary drastic reforms. Rivers or sewers? The choice should be ours.

CHAPTER 11

River Management – Which Way Forward?

The previous chapter has described how river environments are subject to very heavy pressures. These demands continue to increase, and authorities charged with the management of rivers have the daunting task, not only of resolving the conflict of values among the various human exploiters, but also of meeting the responsibility to consider the needs of the natural communities. If rivers are to be more than avenues of navigation, sources of water supply and power and drains to carry away civilization's debris, then built into management schemes should be the concept that rivers are environments worth cherishing. While fulfilling mankind's consumer needs, they may 'also afford beauty and pleasure to man while providing a great resource for wildlife' (Nature Conservancy Council). To achieve this balance, a thorough knowledge of the behaviour and ecological requirements of riverine inhabitants and the human exploiters is necessary, as well as the understanding of how these habitats will respond to their utilization. If river authorities work in close co-operation with scientists and environmentalists then sympathetic though viable management alternatives acceptable to all may well be determined – at least for some rivers. Others, in their present condition, need urgent attention and exceptional treatment. 'Rivers, like people, get sick through taking in toxic substances. The cure is the same too, rest and the right treatment; the longer the "illness", the more difficult the treatment becomes.' (Celia Kirby.)

Water Management

The provision of water services and management is undertaken by a number of authorities. In England and Wales there are ten Regional Water Authorities responsible for water resource development: water supply, sewerage and sewage disposal, pollution control, land drainage, fisheries, navigation, recreation, and river conservation. Private companies are also employed to cater for water supply in some areas. Each of the regional authorities is based on entire river catchment areas which cut across the administrative boundaries of local and national government agencies. In Scotland, the responsibilities are more widely shared. Twelve local government authorities aided by the Central Scotland Water Development Board are in charge of sewerage services and fisheries, while ten river purification boards exercise all remaining functions, but for land drainage. In this area the Scottish Development Department, with grant aid from the Department of Agriculture and Fisheries in Scotland, takes charge. Some responsibilities are discharged through local authorities and/or district councils acting as agents for the water authorities in all three countries. These duties are largely to ascertain the sufficiency and 'wholesomeness' of the supplies in their respective areas. Ministerial responsibility for water rests with the government's Department of the Environment (DoE) for supply and disposal; the Ministry of Agriculture takes charge of land drainage and flood control.

Gouthwaite Reservoir in upper Nidderdale has become an important sanctuary for waterfowl

Pure or Poisoned?

The most important aspect of river management is considered to be the careful control of water quality. This is carried out, theoretically, by regularly sampling water and testing for impurities and, depending on the concentration of pollutants, regulating the quality of discharges into rivers. Successful control of water quality is vital, for, ultimately, much of it returns via the river, after being discharged into water-courses from treatment works, to our taps: there is evidence that high nitrate levels cause blue baby disease and stomach cancer (see page 37), while if sewage were inadequately treated there would be cause for concern, with a whole range of waterborne diseases perhaps becoming rampant.

In theory, rivers should be able to cleanse themselves and purify their waters because of the presence of dissolved oxygen, which activates detritus feeders to feed on most of the polluting substances. But excessive concentration of extraneous substances rapidly lowers the oxygen content (see page 195), reducing the effectiveness of the detritus feeders, and as the decomposing load builds up the whole system spirals down into foul stagnation. The DoE has devised its own classification of rivers based on the oxygen content: see opposite.

The BOD is the depletion of oxygen in a solution brought about by the biochemical breakdown of organic matter by the river's micro-organic inhabitants. It would be desirable if discharge were regulated with consideration of the class of river concerned. The higher the concentration of effluent, the higher will be the ratio of clean water

Classification of Rivers – as Defined by the Department of the Environment

---

Class 1 Unpolluted or recovered from pollution.
Biological oxygen demand (BOD) generally less than three milligrams per litre and well oxygenated.

Class 2 Doubtful quality.
Reduced oxygen content, containing turbid or toxic discharges but not seriously affected.

Class 3 Poor quality.
Dissolved oxygen below fifty per cent saturation for considerable periods, occasionally toxic and changed in character by the discharge of solids.

Class 4 Grossly polluted.
BOD greater than twelve, incapable of supporting fish life, smelling and offensive in appearance.

---

required for cleansing. Where such dilution is not possible, more stringent effluent standards are necessary, but this is not always practically feasible, as revealed in the Tyneside example (page 194).

The state of the nation's sewer networks cannot help the problem. Many are very old and not only need renovating or replacing but, with their small diameter pipes, were not designed to cope with the load and content of modern effluents, and are therefore prone to damage. Although those built before 1920 are dual pipe networks, carrying foul water from the main drains into the sewage works, and surface water from storm drainage separately, directly into the river, fifteen per cent of the nation's networks are over a hundred years old. They were designed to collect, in the same pipe, both rain and foul water, thus overloading the effluent needing treatment.

The cost of new construction, replacing and maintaining sewage networks is prohibitive. In 1983 the cost of nothing more than the maintenance of the present systems in their existing state was calculated at £310 million a year. Yet fifty per cent of the total funds available for waste water treatment have to go toward just the disposal of sewage sludge, and if the treatment plants handle toxic industrial effluent then disposal has to be carried out most carefully. And so it is not surprising that the renewal of the nation's sewerage systems has been urged as a useful means of combating unemployment by numerous politicians.

Toxic industrial chemical effluent is difficult, hazardous and expensive to treat. As yet there is no agreement about where such effluent treatment should take place. Some of it is treated at sewerage works on the basis of a fee charged by water authorities. But when special equipment is required to extract toxic substances, then a factory may be required to install its own treatment plant. Water authorities can maintain strict checks on the effluent leaving sewerage works in their control, but while they can insist on the same high standards from factory treatment works, opportunities to monitor these are limited. Private effluent treatment has to be considered as a part of production costs, but as industry is forced to pay for both the intake of water as well as the effluent discharge, costs can be formidable. Any saving made in cutting back the intake of water is only likely to make the effluent more concentrated, and therefore more costly to treat. It is hardly surprising, therefore, that on several occasions unscrupulous firms have chosen to ignore their responsibilities to the community and have simply tipped their toxic effluents directly into the river. It has been suggested that these circumstances warrant a case for

Water regulation can also threaten wildlife: the coot's nest will be affected by sudden changes in water-level

some kind of rebate if industrial water leaves a factory's premises at, or above, a fixed desirable minimum quality.

Waste disposal firms now offer to cater for the disposal of toxic effluents, but while industrial waste increases, unused 'safe sites', such as disused quarries and sand and gravel pits, are diminishing in numbers, and alternative outlets are no longer easily found. Careless dumping has caused ground water and river reservoirs to become contaminated, and as a result, water authorities are having to impose strict conditions over the licensing of such sites. The EEC now also proposes tighter controls on the application of sewage sludge to land to safeguard against the growing health hazards from metal contamination. Any hope in resolving this nightmarish problem lies in the research that slowly continues, as funds allow, into the improvement of effluent treatment techniques and ways to recycle water and wastes more efficiently within industrial plants.

The Thames is a notable example of a river which was formerly so polluted that it was in danger of posing a threat to public health. It has now been transformed into a river inhabited by many fish and birds showing that man *can* remedy serious river pollution if a clean-up campaign is firmly established. For centuries the river was treated as a rubbish tip into which all manner of wastes were thrown, and with the introduction of water closets the discharge of untreated sewage made the watercourse into an open sewer, together with the rubbish tipped in from the city's 150 slaughter-houses, fish markets and tanneries. And, as the river is tidal, the same foul water flowed back and forth on the tides for days, or even weeks, the estuary thus being prone to stench and stagnation.

Rivers offer many opportunities for recreation, although boating and angling are sometimes conflicting interests. These canoeists are operating on one of the most exclusive sections of the Herefordshire Wye

In the nineteenth century industrial pollution contributed further to the deterioration of the river, particularly from the numerous gas-making plants which were scattered along the Thames. In the production of gas from coal, coal-tar and ammoniacal liquor were removed by passing gas through water, producing a liquid residue of extreme toxicity. Later, when lime was used as a purifier, much of the spent lime found its way into the river.

In the early 1880s a Royal Commission recommended that the discharge of untreated sewage should be discontinued, and from 1889 to 1915 a treatment of lime and ferrous sulphate was used to settle the sludge out of the sewage. The sludge was then taken by ship and dumped at sea. This chemical treatment and an extension of the sewer network began to have a cleansing effect on the river. The Second World War soon put a stop to this process. Pollution grew with bombing and destruction and there were no funds to repair the damaged networks. The post-war period added to this burden. With the growing industrial concentration in and around the city there was an upsurge in London's population, but this was not matched by a comparable extension of the sewage network. The developing suburbs were not linked to the city's main drainage systems, but were instead connected to numerous independent plants sited here and there on the river's tributary streams. These plants were able to perform only the most basic functions and were therefore discharging vast volumes of effluent of very low purity. As a result, pollution soon spread widely throughout the natural network of the Thames waterways – and again the condition of the

The Thames: a Success Story

In the valley of the most well known of all lowland rivers, a study of bankside habitats can be made not only along several stretches of the river itself or its reservoirs, but also at its feeding streams such as the R Crane and R Brent on the N bank, or the Wandle and Pool on the S.

In Greater London, some sites showing interesting riverine habitats include the following.

Brent Reservoir, NW9. The silted areas where R Brent and Silk Streams enter the reservoir exhibit a succession from open water to willow carr. This is a wintering site for gadwall.

Lavender Pond, SE19. A Thames-water artificial pond with alder carr and flood meadow communities. Many water birds and at least four species of dragonfly.

Ruislip (RDNHS). The reservoir offers many habitats in open water and the surrounding marshland. The reserve is fringed by oak and coppiced hornbeam woods and there are therefore many woodland birds in the area. Birds include woodcock, water-rail, whitethroat, hawfinch, wood warbler and tree pipit.

Staines Reservoirs, Hanworth. Famous for their large winter concentrations of wildfowl including pochard, tufted duck, teal and wigeon.

Ten Acre Wood, Hayes. Old hazel-coppiced woodland, through which Yeading Brook flows.

Walthamstow Marsh (Lea Valley Authority), Clapton. Fen-type communities. Birds include sedge warbler, snipe and teal. The marsh joins *Walthamstow Reservoirs* (Thames Water Authority), famous for its heronry. Other resident and migrant species include mute swan, Canada goose, greenshank, redshank, snipe, dunlin, green sandpiper, kingfisher, pied and yellow wagtail.

Perch, an attractive but carnivorous fish, is sometimes removed in angling waters by electrofishing

river became a major threat. 'Suicides from London's bridges were not so much drowned as poisoned, and stomach pumps were part of the standard rescue equipment.' (L. M. Bates, 1977.)

In 1959 another anti-pollution campaign was activated, beginning immediately with the rebuilding and extension of the sewage network and associated treatment and aeration plants. Under normal conditions the network can now handle all the domestic and industrial waste arriving at its treatment plants, but it is still beyond its capabilities to cope with the large amounts of surface run-off during the occasional freak storms when sewers overflow into storm water channels carrying untreated sewage into the river. Resulting deoxygenation causes fish mortalities: in 1973 thousands of fish died in the Wandsworth Reach following torrential rain and consequent storm water discharges.

Stringent by-laws control the discharge of

Some alien fish have been introduced to British rivers. The small bitterling *a* is little problem, but the voracious zander *b* is a controversial introduction

pollutants by riverside industry. Industrial concerns have been forced to improve their private treatment plants, and where large volumes of effluent are discharged, factories are encouraged to install aerators to raise the level of dissolved oxygen. But the problem of hundreds of tons of rubbish – plastic packaging, bottles, cans, rope and balks of timber – that have to be collected from the river each year still has to be solved. Since 1960 new power stations have been built downstream where the larger volumes of water are able to absorb the excess heat with

a less serious effects. All these measures have helped to stop the depletion of the river's oxygen content, which is now constantly monitored.

Fish began to return in the 1960s when the river's waters became hospitable enough to support an abundant invertebrate fauna, and by the end of 1975 eighty-six species of freshwater and marine fish were reported to have been caught on the water intake screens of power stations. Fish and other water creatures began attracting birds, and today the twenty-five mile stretch between London Bridge and Tilbury is home for a large number of overwintering wildfowl and waders. The Thames is now rated as the world's cleanest urban estuary!

Monitoring the Water

Water for consumption can theoretically be collected from any point on its journey to the sea, but as this supply varies through the year, careful regulation is necessary. The amount of water that can be abstracted to meet consumer demand depends on how much needs to be left behind so as not to upset the biological regime. There must, for reasons already described, be adequate

b quantities for dilution of the effluent from sewage works. Where an abstraction point is situated close to an estuary, freshwater flow falling below what is considered to be the safe level will gradually increase the saline content of the river water upstream, and so interfere with the quality of water at the abstraction point. Industry sited close to abstraction points will also affect the required minimum flow to dilute effluents adequately.

Drinking water is abstracted at specially constructed sites where the river has been ponded by weirs. Here the water has been treated to the highest qualities. In these times of dangerous industrial effluents water authorities must plan to ensure that there are always sufficient supplies of stored drinkable water should abstraction have to be ceased

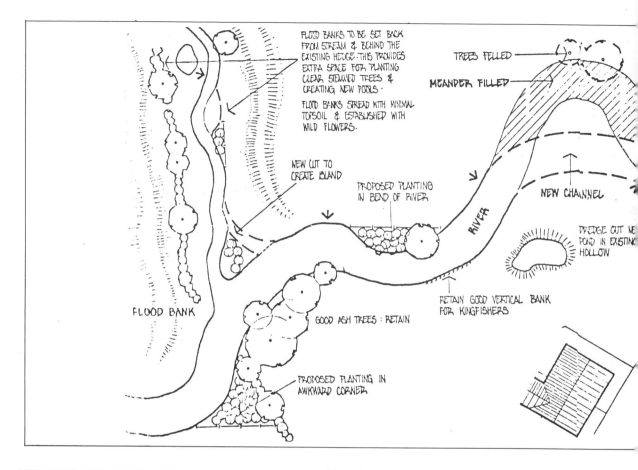

FLOOD BANKS TO BE SET BACK FROM STREAM & BEHIND THE EXISTING HEDGE. THIS PROVIDES EXTRA SPACE FOR PLANTING CLEAR STEMMED TREES & CREATING NEW POOLS.

FLOOD BANKS SPREAD WITH MINIMAL TOPSOIL & ESTABLISHED WITH WILD FLOWERS.

NEW CUT TO CREATE ISLAND

PROPOSED PLANTING IN BEND OF RIVER

TREES FELLED

MEANDER FILLED

NEW CHANNEL

RIVER

DREDGE OUT NE POND IN EXISTIN HOLLOW

RETAIN GOOD VERTICAL BANK FOR KINGFISHERS

FLOOD BANK

GOOD ASH TREES : RETAIN

PROPOSED PLANTING IN AWKWARD CORNER

Rutland Water (L and RTNC-AWA), Leicestershire and Rutland, SK 897049. SE of Oakham. This is the second largest man-made lake in England, lying close to bird migration routes along the Welland and Nene rivers. It has become of national importance because of its wintering wildfowl. Careful planning and excellent management encourage wildlife. Shallow lagoons have been created by building banks which may be flooded when the water is high, but which hold their levels when the reservoir draws low. The levels of water can also be individually regulated, so that species feeding at different depths can be accommodated as and when they flock here. As it is a recreation reservoir, it has been zoned to give wildlife maximum sanctuary. 210 bird species have been recorded: some nesting in the grassland around the reserve, others being passage migrants.

suddenly in the event of unforeseen pollution, such as a spillage of toxic wastes from a local factory.

With ever increasing consumption and the problems brought on by seasonal periods of drought in recent years, the natural flow of rivers in some parts of Britain has been inadequate. Recent suggestions of vast national networks of inter-linked river systems may solve the problem of supply, but the mixture of the water's physical and biological contents may affect plant and animal populations dependent on a particular nutrient supply.

Reservoirs are another solution. The latest constructions are being designed to be a part of a whole river catchment basin where natural river channels are used to distribute water, with the effect that dry weather flows are now double their natural levels. But

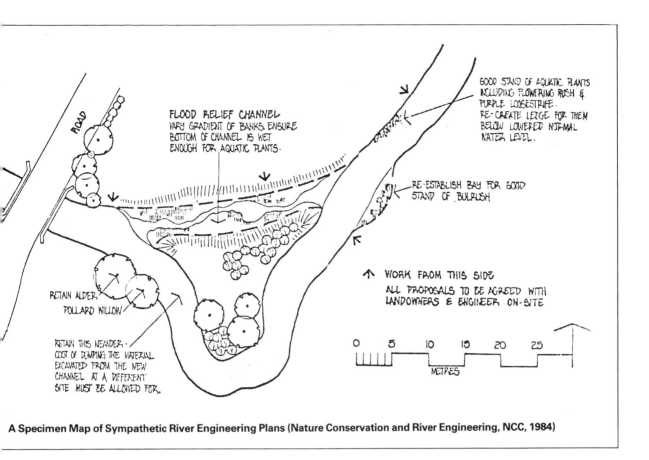

A Specimen Map of Sympathetic River Engineering Plans (Nature Conservation and River Engineering, NCC, 1984)

there are limits to the size of reservoirs and the amount of stored water supply. So it is important to improve the quality of effluent discharge further downstream, enabling abstraction from lower volumes of fresh water instead of continuing to increase volumes with storage water simply to make the biological quality acceptable. The gauge for the required volume of water to be released from regulatory reservoirs is based on the level of the abstraction point furthest downstream. At this point a predetermined 'minimum acceptable flow' level is fixed and then enough stored water is released to maintain the volume of the river above this level. Examples of rivers which are regulated in this way are the Dee, the Severn and the Tees.

Under conditions of drought, when there is a shortage of water supply – such as in 1975 and 1976 when even Parliamentary intervention was necessary to impose restriction on the use of water – minimum acceptable flow levels may have to be reduced. In contrast, during the wet season, allowances have to be made for flood alleviation. Autumn levels have to be deliberately reduced to leave room for winter floodwater. But ingenuity and foresight are required to get the balance right, for if a winter is dry then supplies for subsequent months could be dangerously low.

Fishery Management

The regulation of the quantity and quality of fish stocks is another aspect of river management. In some rivers famous for their game fishing, unwanted species, such as pike and perch, are removed by

The Evaluation of Rivers for Nature Conservation and Management
Rivers have been categorized for the purposes of nature conservation and management. This is evaluated on the basis of the geology, chemistry and geography of the area through which it flows, its size and extent, the rarity of its flora and fauna, the 'naturalness' and diversity of its habitats, and its fragility.

Nature Conservation and River Engineering NCC 1984

electrofishing – alternating current is used to stun the fish so that they can easily be gathered up with a hand net. The rivers are then restocked with game fish such as rainbow trout or brown trout reared in hatcheries. Hatcheries usually choose sites where cool, clean water is in good supply, such as near natural springs. Fish do grow faster in warm water, but they are more vulnerable to fungal and bacterial diseases. However, if biological conditions for rearing fish are otherwise right, some hatcheries have chosen sites close to power stations, to take advantage of the warm water released by the cooling towers, which is then treated for purity and artificially oxygenated. Carp in particular does very well in such waters. A high flow is also desirable to sweep away organic wastes created by large concentrations of fish which would otherwise lower the oxygen content of the water. Again vast flushes released from cooling towers serve this purpose.

Man's increasing control over water flow through the construction of reservoirs and dams for hydroelectric power and river flow regulation has taken a toll on the natural stocks of the salmonoids. These barriers and disturbances have had the effect of interrupting their migratory paths up-stream. Attempts have been made to amend this situation. The Scottish River Tay is one of Britain's most valuable salmon rivers and when the hydroelectric scheme of Pitlochry dam was built, a 1000 foot long fish ladder with thirty-four pools was constructed. In this way the salmon cover a fifty foot rise by

moving through the pools and interconnecting pipes. Visitors to the dam are able to watch the migrating fish through one of the glass-sided pools.

Elsewhere research is being carried out on another aspect of migratory fish. In 1983 the South West Water Authority and the Fisheries Laboratory of the Ministry for Agriculture, Fisheries and Food at Lowestoft joined forces to investigate the salmonid movements in the River Towey in Cornwall prior to a large reservoir being constructed at the river's headwaters. The Towey is well known for its salmon and sea-trout runs. Salmonid runs are understood to be signalled by a surge in

Height of source	Geology	Nutrient status	River and type of vegetation
<200m	Limestone		Durness streams
	Chalk	Rich	Test ○○○○○○○○● Kennet ○○○○●●● Mimram ○○○○
		Moderate	Itchen ○○○○○○ Piddle ○○○○○● Hull ○○○○●
	Intermediate	Rich	Loddon ○○●●●● Bure ○○○●●●●○ Stour ○○●●●●●●
		Moderate	Frome ○○○○○○●● Wensum ○○ Avon ○●●●●●●●●●●●●
	Clays, Gravels Sand	Rich	Eye ○○●●●●● Beult ●●●●● Thame ○●●●●●●●●
		Moderate	Torridge △▲■▲■■■■■■ Ugie △△△○○○ Gwendraeth ■■■■
		Poor	Carron ☆☆☆★ Highland ☆△ Blackwater ☆△
200m-500m	Sandstone or Limestone	Rich	Tweed ★△△△△○○□□○○○○○○○○○○○ Windrush ○○○○○●●●●●
			Lugg △■■■■●●○●●
		Moderate	Teme ▲△■■■■□■■■■■■● Tamar △△■■■■■△■■
			Dove ★■■■■■■■●■
		Poor	Horner ☆▲▲■ Craigroy ★★▲
	Resistant rock	Rich	Endrick ★△▲△○○ Ythan △○○□□□○ Dane ★△△○●●●
		Moderate	Spey ☆☆☆△☆△△△△△△ Granneli △△△ Erme ☆☆△○
		Poor	Oykel ★★★★★ Cothi ☆☆△△▲▲▲▲ Conway ☆☆▲▲▲△
>500m	Sandstone or Limestone	Rich	Monnow △■■■■■■■ Ribble ★△■■■○■■○○■■■○
			W. Dee ★△△△△△△△△△■○●●●●
		Moderate	Usk ▲■■■■■■■■■■□ Earn ▲▲△○○□
		Poor	Balnagown ★▲▲
	Resistant rock	Rich	Unnatural
		Moderate	Teifi ☆△△△△△△○△▲■■■■ Don ★★△△△△△△△○○○○○○□
			Cocker ★△△△
		Poor	S. Dee ☆☆★★★★★☆△△△△△△ Findhorn ★★★★★△☆★▲▲▲▲△△
			Derwent ☆△△△▲△□▲

Communities of plants found along a selection of British rivers. The symbols show different vegetation types associated with varying physical characteristics and a downstream succession from left to right.

Key to river types
○ Typical chalk stream community
● Typical of lowland clay rivers
□ Community associated with sandstone and limestone
■ Community associated with mixed sandstones and limestones
△ Typical community of nutrient poor shales
▲ Typical community of more enriched shales

☆ Community associated with highland streams with flowering plants and bryophytes
★ Community associated with bryophyte dominated highland streams

water flow. It is thought that a natural 'freshet' (an increase in the volume of water that may occur after snow-melt or heavy rains) will stimulate upriver migration. Tests are being carried out to observe whether artificial freshets, such as those caused by stored water released from reservoirs, can be as effective as natural floodwaters in stimulating a response. Fish are to be tracked by radio-pills inserted into their stomachs. In addition, information is being sought on the exact minimum water flows required for migratory fish, because engineers consider that stored water released simply because volumes are low in migratory rivers is 'wasted water'. They have pointed out that if waters released in England and Wales could be reduced by about twelve per cent then there would be a saving of about £100 million in the required reservoir capacity. Better management may emerge from such research, not only to seek the survival of rivers used by migrating salmon on less costly, leaner volumes, but also for the kind of conditions that will lead to better natural stocks of salmonids and a higher rate of surviving and returning fish.

Drainage or Damage?

Flood alleviation schemes which improve farm drainage and reduce the flooding of property feature large in river management.

Most readers will recall the vivid pictures of flooding in the winters of recent years when property was damaged and life threatened. Further, successive governments have pushed for greater food production – and encouragement towards undertaking wide-scale land drainage has come in the form of enormous grants given to farmers and drainage authorities. The way to solve the problem of flooding in many areas, shown by the legacy of the massive scars that criss-cross East Anglia and the Midlands, has been to cleave out geometrical channels with heavy machinery, with no regard for the marginal habitats, trees and wildlife inhabitants.

But it is possible (as is described in the NCC's guide entitled *Nature Conservation and River Engineering*) to carry out effective river engineering for the purpose of drainage and, at perhaps no more than a little additional cost, preserve and create viable riverine habitats. Pages 214/215 show a specimen map of such a river engineering scheme, where successful results depend on knowledge, sensitivity and specialisms which embrace ecology and landscaping, and also on engineering techniques that require special skills, such as being able to cut meandering channels within inches of important plants or marginal spawning shelves, or working from only a single bank so that stretches of wildlife can be left untouched and undisturbed. Since water authorities who carry out this work seldom own any land alongside rivers, landowners' approval has to be sought to activate such schemes. But it seems as though river authorities have found that farmers will concede land along a river so that flood-

banks can be created to form a variety of angles of bank slopes, making for more marginal features such as ledges, cliffs and steep banks, bays, pools and ditches. Such a pioneering scheme, carried out by the Severn Trent Authority with the involvement of naturalists, environmental groups and landowners, has found that recently created marginal riverine habitats were already supporting several species of dragonfly and a host of aquatic animals and plants as well as a rising population of water birds.

The NCC has a procedure for grading the relative values of British rivers which is based on a scale of habitat and wildlife interest within the Council's own guidelines. The guidelines, which cover all habitats across the range of British river types, reflecting all physical and chemical variation (pages 216/217), are continuously updated as more is known about the inhabiting species and communities. At the top end of the scale are Sites of Special Scientific Interest (SSSI), important to nature conservation, including riverine sites of National Nature Reserve (NNR) status. In all but a handful of these rivers the interests of nature conservation feature high and the aspects of 'management' that we have considered so far are in fact 'disturbances'. Therefore management here is minimal or is determined by the levels acceptable to sustain the wildlife of these river systems. But minimum standards are set for all river types and the NCC's strategy for those falling below SSSI standards 'is one of persuasion or advice' to the concerned statutory bodies or individuals to minimize the continued losses of wildlife and habitats

Opposite above An aerial view of the Severn in flood, in Gloucestershire. Flood alleviation schemes are an important part of river management and the Severn Trent Water Authority is operating a pioneering scheme incorporating ecology and conservation into river engineering operations *(Cambridge University Collection)*

Opposite below Upper Donside in north-east Scotland. Unlike many Scottish rivers, the Don is not affected by hydroelectric power schemes although downstream towards Aberdeen a series of paper mills have exploited the rivers

within the rivers for which they are responsible.

Reviving Tradition

Traditionally, fen habitats were maintained under an established regime of reed cutting, mowing and grazing, and it may still be economically viable to continue these practices in order to retain the characteristic plants of the area. Basket willows, *Salix viminalis* or 'osiers', were used to make many formerly useful articles of commerce: bullock feeding baskets, bushel skeps, fish baskets, linen baskets and the like. Today in East Anglia, the water authority grows osiers especially for weaving 'mattresses' which are used in river-bank stabilization. Some of these structures have been in place for forty years and still show no signs of needing to be replaced – and they are much more hospitable to wildlife than their concrete or metal sheeting counterparts used elsewhere. Rush species, such as reed mace (bulrush) were commonly employed in the craft of carpet and mat weaving. Reed cutting for thatching is still carried out in Wicken Fen and in Norfolk, where its profits may offset the cost of maintaining the original Fenland.

The most typical plants of the Fens and Broads are the extensive beds of common reed (*Phragmites autralis*) and saw sedge (*Cladium mariscus*) – the nearest nature has ever come to producing monocultures – and these plants actually do thrive on the waters of the Fens and Broads, over enriched as a result of fertilizer run-off. However, if the reed and sedge are cut and mown regularly then the natural pattern of monocultures is changed into polycultures in which many beautiful and delicate herbs and marshland plants find niches in which to grow. Controlled and well-timed grazing has the effect of discouraging coarse meadow grass and, if limited to the winter months, it will not destroy the growth of the shoots and buds of marshland plants and grasses in

The lowering of water-levels can lead to a change to alder and fen carr environments

spring. The activity does not affect overwintering buds. Some waterside plants, such as the rare adder's tongue spearwort, will in fact flourish in cattle-trampled margins.

Where drainage and intensive farming methods have been drastic they have brought about a considerable lowering of the water-table. Because waterways in river catchment areas can be composed of vast networks of interrelated systems, water-tables well outside the area of drainage may suffer. Local wetlands that were dependent on a high water-level dry out, and their native species disappear unless a high water-level can be artificially maintained. In the Fens, coarser vegetation dominated by shrub and tree species, such as alder and sallow, invades to create fen carr. The cost of artificial flooding can be prohibitive as the water-table is progressively lowered, because more lifting power is needed to get the water up and out. However, in some instances, where old flood relief measures have been maintained, it has been possible to divert water into the Fenlands using them as winter storage areas where, while relieving flood dangers elsewhere, the nutrient-rich waters provide a good, if late, crop of hay, as well as a habitat for many Fenland plants and native and visiting bird species. But it is important that this source of water be little polluted, or some plants will become established at the expense of others, creating unwanted changes in the fen vegetation.

The Ouse Washes in East Anglia, lying between the Old and New Bedford Rivers, were created as a flood relief measure in the seventeenth-century drainage scheme described in Chapter 3. Today they are a site of international importance for wildlife. During the regular winter flooding, the sluices are opened on to the Washes, and the flooded meadows become the feeding and resting grounds for vast numbers of overwintering and migratory birds such as ducks, geese and swans and, most notably, the rare Bewick's swan. The rich summer hay meadows are ideal grazing grounds as well as a breeding place for birds such as the black-tailed godwit and the ruff.

Sanctuaries or Island Cells?

As man's needs and technology change, what was created in one period of history is usually changed or destroyed in the next, and aquatic wildlife either disappears or learns to adapt to new regimes. The niches for wetland wildlife are however becoming increasingly confined to the 'islands' of nature reserves in which a principal aim of the management of rivers, as the NCC has pointed out, is to create minimal environmental impact while maximizing the number of habitats. But one nature reserve may frequently be isolated from the next sanctuary by vast seas of modern farmland or even conurbations. While this creates problems of dispersal for certain species, these isolated reserves are vulnerable to the management of land around them, which will affect organic pollution and water-levels. Wicken Fen, a nature reserve owned by the National Trust, is a national sanctuary of great importance. It has

> *Ouse Washes* (Cambient, RSPB and WT). Cambridgeshire and Norfolk.
> Winter flooded fen meadows. Although it is birds that the Washes are visited for, particularly the winter waterfowl, the meadows are also rich in plant and insect life, including some uncommon species such as the large tortoiseshell butterfly and obscure wainscot and cream-bordered green pea moths. The rivers are famous for large rudd and pike and also include bream, perch, tench, eel and zander. Many mammals also inhabit the meadows. *The Welney Wildfowl Refuge* (WT), TL 547946, in Norfolk, is part of the Washes. N of Littleport off A1101.

contributed enormously to our knowledge of Fenland ecosystems. Of great interest to entomologists and botanists, it has some hundreds of species of flowering plants and moth and butterfly and thousands of insect species. However, due to drainage and peat shrinkage in the surrounding area the water-level has dropped, and to maintain the reserve, water has now to be pumped up into it.

Only four per cent of land in Britain is safeguarded for nature conservation. But even within this low percentage are areas of conflict – and yet the NCC considers that if we are to rely on nature reserves as our means of conserving our flora and fauna then we need a great many *more* areas for this purpose. This is quite unlikely to happen, particularly in the more productive low-lying farmlands of Britain, while the charm of British wildlife has always been inseparable from the countryside of which it is a part.

The system of National Parks and Areas of Outstanding Natural Beauty in England and Wales aims to conserve some of our finest tracts of countryside, while at the same time allowing people to live and work in them – but these areas are concentrated on the uplands, incorporating only certain types of countryside, such as moors or

The Ouse Washes provide a vital sanctuary for wildfowl

mountains, or they comprise stretches of coastal scenery. There are no representative large expanses of fen country or coastal marshland amongst which lie some of our most beautiful wetland and riverine tracts, such as the Norfolk Broads, the Halvergate Marshes, the Somerset Levels and the Exe valley in Devon, although some smaller stretches do carry SSSI or NNR status. These wetlands are being gravely threatened, and yet, because of the political influence of powerful vested interests, none has been given the protection of National Park status. One of the problems is that access to these areas is largely confined to public rights of way, in contrast to the more general freedom to wander which may exist – or be thought to exist – in upland areas.

A Successful Alternative

One of the problems of confining wildlife in special sanctuaries like nature reserves is that it has less immediacy to people living in cities, who have to travel several miles to enjoy 'unspoilt' river meadows and marshland habitats. Imaginative planning and creative management have led to a few surprisingly successful solutions, as at Peterborough in Cambridgeshire. When Peterborough was designated a New Town in 1967, a Development Corporation was set up with the aim of planning expansion with care, so as to limit the destruction of the surrounding countryside. An important aspect of the New Town is a valuable haven for wildlife in the form of Nene Park. The Royal Society for Nature Conservation and the Countryside Commission were involved to represent the interests of nature conservation from the initial planning stages. The Park covers over 2000 acres of undeveloped land along the River Nene. Its character changes subtly from east to west – the general concept being 'that it should flow from the urban centre through informal, open areas to open countryside at the western end'. Ferry Meadows is the centrepiece of Nene Park, 120 acres of man-made lakes, interconnected and linked to the River Nene. Although the meadows offer leisure activities such as sailing and

Sometimes a riverside can be incorporated into a suburban park, as at Peterborough, or here, on the Avon at Salisbury

fishing, the area is said to be 'an exciting place for wildlife' and is a centre for Naturalists' Trust meetings and excursions. The RSNC officer, Sarah Douglas, claims that her job 'is mainly with the non-expert (naturalists) convincing them that they want to find out more about the wildlife in the area'. As a result, Ferry Meadows is proving to be extremely successful, since it combines leisure activities with nature conservation and education and is thus an enjoyable place to visit. This venture of taking conservation into an urban vicinity may encourage people, made aware of the ecology of the water habitats on their doorstep, to take a concern in those elsewhere.

If we are to hope for the future of our rivers and interlinked wetlands in this highly populated and intensely cultivated country then responsible management offers the only answer. For most rivers, management is required almost continually. The ever increasing domestic, agricultural and industrial effluent needs better sewerage systems and water treatment plants and the maximum possible amount of recycling, but these are costly and slow to appear. Flood alleviation and land drainage schemes can be applied, where possible, in the commonsense ways described by the NCC and, with the co-operation of farmers, water authorities and environmental groups, watercourses can become more than just drainage channels. Often it is a case of taking just that little more trouble in the planning stages rather than spending a lot more money putting things right.

Success stories, like those of the Thames and Nene Park, deserve to be praised, so long as the acclamation does not divert attention from the fact that the destruction of the general national environment continues at a terrifying rate. It is hard to avoid the conclusion that, were the people of Britain invited to take the important decisions, most problems would be solved. The cost of enhancing and protecting the natural beauties and wildlife of our waterways is modest in comparison to the benefits. Yet one cannot avoid the feeling that while past generations of schoolchildren could look forward to their nature walks, future generations may only encounter nature in the museum, never knowing the delights of wholesome, unvandalized countryside.

Index